R. Klatte   U. Kulisch   M. Neaga
D. Ratz   Ch. Ullrich

PASCAL-XSC

Language Reference with Examples

Translated by
G. F. Corliss   R. Klatte   U. Kulisch
D. Ratz   C. Wolff

Springer-Verlag

Berlin Heidelberg New York
London Paris Tokyo
Hong Kong Barcelona
Budapest

Prof. Dr. Ulrich Kulisch
Dr. Rudi Klatte
Dipl.-Math. techn. Dietmar Ratz
Dipl.-Übers. Carola Wolff
Institut für Angewandte Mathematik
Universität Karlsruhe
Kaiserstraße 12
W-7500 Karlsruhe, FRG

Prof. George F. Corliss
Mathematics and Computer Science
Division
Argonne National Laboratory
9700 S. Cass Ave.
Argonne, IL 60439-4844, USA

Dr. Michael Neaga
Numerik Software GmbH
P. O. Box 2232
W-7570 Baden-Baden, FRG

Prof. Dr. Christian Ullrich
Institut für Informatik
Universität Basel
Mittlere Straße 142
CH-4056 Basel

Title of the original German edition:
PASCAL-XSC Sprachbeschreibung mit Beispielen
© Springer-Verlag Berlin Heidelberg 1991

Mathematics Subject Classification (1991): 65-XX, 65-04, 65G-XX, 65G-05, 65G-10, 68-XX, 68B99

ISBN-13:978-3-540-55137-9   e-ISBN-13:978-3-642-77277-1
DOI: 10.1007/978-3-642-77277-1

Typesetting: Camera ready by authors

41/3140-543210 - Printed on acid-free paper

Preface

This manual describes a PASCAL extension for scientific computation with the short title PASCAL–XSC (PASCAL eXtension for Scientific Computation). The language is the result of a long term effort of members of the Institute for Applied Mathematics of Karlsruhe University and several associated scientists. PASCAL–XSC is intended to make the computer more powerful arithmetically than usual. It makes the computer look like a vector processor to the programmer by providing the vector/matrix operations in a natural form with array data types and the usual operator symbols. Programming of algorithms is thus brought considerably closer to the usual mathematical notation. As an additional feature in PASCAL–XSC, all predefined operators for real and complex numbers and intervals, vectors, matrices, and so on, deliver an answer that differs from the exact result by at most one rounding.

Numerical mathematics has devised algorithms that deliver highly accurate and automatically verified results by applying mathematical fixed point theorems. That is, these computations carry their own accuracy control. However, their implementation requires arithmetic and programming tools that have not been available previously. The development of PASCAL–XSC has been aimed at providing these tools within the PASCAL setting.

Work on the subject began during the 1960's with the development of a general theory of computer arithmetic. At first, new algorithms for the realization of the arithmetic operations had to be developed and implemented. The design and development of appropriate programming languages began around 1975 with preliminary implementation studies first on the basis of PASCAL and also as an extension of FORTRAN. As the next step, complete compilers for the extended language had to be developed. Since about 1980, algorithms for standard problems of numerical analysis with automatic result verification and for many applications have been systematically developed.

Many colleagues and scientists closely related with the Institute have contributed to the project by useful discussions, by a long term collaboration, or other kinds of support. The main participants of this developement are: U. Allendörfer, J. H. Bleher, H. Böhm, G. Bohlender, K. Braune, D. M. Claudio, D. Cordes, G. F. Corliss, A. Davidenkoff, H. C. Fischer, S. Geörg, K. Grüner, R. Hammer, E. Kaucher, R. Kelch, R. Kirchner, R. Klatte, W. Klein, W. Krämer, U. Kulisch, R. Lohner, M. Metzger, W. L. Miranker, M. Neaga, L. B. Rall, D. Ratz, S. M. Rump, R. Saier, L. Schmidt, G. Schumacher, D. Shiriaev, Ch. Ullrich, W. Walter, M. Weichelt, H. W. Wippermann, and J. Wolff von Gudenberg. The authors would like to express sincere and cordial thanks to each one for his cooperation. Thanks are also due to the many students who used and applied PASCAL–XSC in an early stage of the development and thus helped to stabilize both the language and the compiler.

This manual provides a complete description of PASCAL–XSC. The part dealing with ISO Standard PASCAL is only briefly discussed, while the extensions marked by [‾‾‾‾‾‾‾PASCAL–XSC] are presented in full detail. A detailed chapter with exercises and solutions is included in this manual to help the reader to get familiar with the new language constructs. A full set of syntax diagrams, appendices, and indices complete the book.

Finally, we would like to mention that a programming language is never complete. Improvements are always possible and often necessary. The main concern developing this language was to provide a useful and appropriate tool for numerical applications in the field of engineering and scientific computation. Benevolent and critical comments for improvements of the language are very welcome.

This book is a translation of a German version also published by Springer-Verlag. The authors are very grateful to George Corliss who helped to polish the text and the contents.

Karlsruhe, October 1991   The Authors

The Realization of this Book

This manual was completely written in the text system LaTeX or TeX. Co-Author Dietmar Ratz gathered the text, designed the necessary *macros* and *environments*, developed the syntax diagrams, carried out corrections, and drew up the final version including the appendices and indices. He was also responsible for the final layout of this book.

The Authors

Table of Contents

Chapter 1

Introduction

This book describes the language PASCAL–XSC. The core of the language description consists of three chapters: *language description, arithmetic modules,* and *exercises.*

In chapter 1 (Introduction), we describe the notation used in this book. We sketch the historical development of PASCAL–XSC, the axiomatic definition of computer arithmetic, and its embedding in programming languages. The last section is a short survey of the language PASCAL–XSC.

Chapter 2 (Language Reference) comprises the formal language definition. The ISO PASCAL Standard is only touched upon. The extensions of PASCAL–XSC are described in detail.

PASCAL–XSC supports arithmetic on real, complex, interval, or complex interval objects, as well as on vectors and matrices over these types. Chapter 3 (The Arithmetic Modules) describes the modules supporting these types with their operators, functions, and procedures. The succeeding chapter 4 (Problem Solving Routines) summarizes the routines which have been developed in PASCAL–XSC for solving frequently occurring numerical problems.

The closing chapter 5 (Exercises with Solutions) encourages the reader to apply the new language elements to easy exercises to extend his or her knowledge. Solutions are provided.

Finally, the Appendix contains the syntax diagrams of PASCAL–XSC, as well as complete lists of reserved words, predefined identifiers, operators, functions, and procedures of the language core and the arithmetic modules.

This book does not deal with implementation details of the language. For all implementation-dependencies in the following chapters, we refer to the corresponding user manual supplied with the special compiler version.

1.1   Typography

To mark or emphasize certain words, names, or paragraphs, we use the following type faces:

italics   serves to emphasize certain words in the text.

boldface   is used to mark reserved words of PASCAL–XSC (e.g. **begin,** **module**) in the text or in excerpts of programs.

slanted   characterizes predefined identifiers of PASCAL–XSC (e.g. *integer*, *real*) and identifiers from programming examples when they appear in the body of the text.

typewriter   is used for listings and run-time outputs of programs that are directly copied from a file or from printer output.

References are always indicated as [*nr*]. The number *nr* corresponds to an entry in the bibliography.

1.2   Historical Remarks and Motivation

In general, electronic computers for engineering and scientific computations are equipped with a floating point arithmetic to approximate mathematical operations with real numbers. All higher programming languages permit these operations to be denoted with the usual operation symbols so that the programmer is able to write down simple expressions, formulas, or functions in their usual mathematical notation. In mathematics and the natural sciences, the concept of the arithmetic operation or function is by no means restricted to real numbers. For example, operations in vector spaces and even vector-valued functions occur. It is not at all efficient to program these concepts on the computer using basic floating point operations and then realize them via clumsy procedure calls, since this may result in many unnecessary computing errors.

Therefore, intensive research in the field of computer arithmetic has been conducted at Professor Kulisch's Institute for Applied Mathematics at the University of Karlsruhe since the 1960's. To achieve satisfactory results in many applications, the computer must support an arithmetic which is much more powerful than the usual floating-point arithmetic. That is, every computer, whether large or small, used for scientific computation should be a *vector* computer[1] in the mathematical sense. Its arithmetic should support operators in the common mathematical vector spaces such as real numbers, complex numbers, intervals of real and complex numbers, or vectors and matrices with elements of these types. The results of these operations should be provided with higher accuracy than can be achieved using the usual floating-point arithmetic. By 1976, a complete mathematical analysis of these demands led to the publication of two books ([24], [28]).

To realize these demands, algorithms and fast hardware circuits have been developed and implemented completely. Today, a variety of realizations is available for all kinds of computers, e.g. personal computers, workstations, general-purpose computers, mainframes, as well as supercomputers. The new operations, e.g. the product of two matrices, always deliver a result which differs from the exact result by at most one single rounding in each component. Assuming that the same technology (software, microcode, hardware, pipeline technique) is used, the new operations

---

[1]The term vector processor is often used as a synonym for a computer that is equipped with pipeline operations. Here, we do not mean this. The concept is used in a more mathematical sense.

are not only more accurate, but also faster in general than those simulated via the traditional floating point arithmetic. Gradually, the manufacturers have realized the correctness and usefulness of this procedure. Thus, over the years, more and more products that support the new demands of arithmetic have appeared on the market.

Immediately, however, difficulties arose concerning the programming languages. Traditional programming languages as ALGOL, FORTRAN, PL/1, PASCAL, or MODULA do not allow access to a hardware-supported matrix product, a multiplication of complex numbers with maximum accuracy, or an interval operation via the traditional operation symbols. Thus, a further development of programming languages was necessary to support the requirements of high quality arithmetic. Between 1976 and 1979, two institutes from the Universities of Karlsruhe and Kaiserslautern (Prof. Kulisch and Prof. Wippermann) cooperated to develop and implement a PASCAL extension called PASCAL-SC (PASCAL for Scientific Computation). In the following years, in cooperation with IBM, a corresponding FORTRAN 77 extension was developed and implemented for IBM/370 computers at the Institute for Applied Mathematics at the University of Karlsruhe. Today, the result is available as an IBM program product under the name of ACRITH-XSC.

ACRITH-XSC contains some constructs such as dynamic arrays and overloading of function names which were not considered in PASCAL-SC. So the language PASCAL-XSC was developed in parallel with the development of ACRITH-XSC. PASCAL-XSC is implemented using a PASCAL-XSC-to-C preprocessor (itself written in C) and a run-time system written in C. Hence, PASCAL-XSC may be installed and used in a nearly identical way on almost any computer system which supports C. In particular, PASCAL-XSC runs on nearly all UNIX systems. Thus, the programmer may develop PASCAL-XSC programs on a personal computer or a workstation and run them on a mainframe or a supercomputer.

Mathematicians have used PASCAL-SC, ACRITH-XSC, and PASCAL-XSC to solve a variety of problems. Easy access to interval operations played a major role. We can use intervals to represent bounds for the solution of the problem or to represent a continuum of the real or complex numbers. A single evaluation of a function over an interval may be sufficient to state in a strict mathematical sense that the function does not possess a zero in this interval. In continuation of these ideas, mathematical fixed point theorems of the Brouwer or Schauder type may be applied to obtain statements on existence and uniqueness concerning numerical problems by the computer itself or to have the correctness of a computed result automatically verified by the computer. Thus, program packages have been developed for the solution of boundary value and eigenvalue problems of ordinary differential equations and systems of linear and nonlinear integral equations. These programs verify the existence as well as uniqueness, and compute narrow bounds for the solution itself (see [27]). The new tools are applied in many fields of application, including mechanics, chemistry, chaos theory, or in the search for periodic solutions of differential equations. Moreover, researchers have been able to discover surprising solutions to problems not previously solvable. For further information, see the References.

1.3   Advanced Computer Arithmetic

When the programming languages ALGOL and FORTRAN were developed in the 1950's, specification of the arithmetic was left to the manufacturer. As a consequence, the programmer does not know what happens when the symbols +, -, *, or / are applied. Besides that, two computers from different vendors may also differ in the properties of arithmetic. Consequently, numerical analysis could not be based on universally valid axioms of computer arithmetic.

The increasing efficiency and speed of computers requires a more precise definition of the arithmetic. Since a computer may represent only a finite set of numbers, the set $I\!R$ of the real numbers has to be mapped onto a subset R, called *floating point numbers*. This mapping $\Box : I\!R \rightarrow R$ is called a rounding if it satisfies the properties:

(R1) $\Box a = a$,   for all $a \in R$, and   (projection)

(R2) $a \leq b \Rightarrow \Box a \leq \Box b$,   for all $a, b \in I\!R$.   (monotonicity)

A rounding possessing the property

(R3) $\Box(-a) = -\Box a$,   for all $a \in I\!R$ (antisymmetry)

is called *antisymmetric*. The commonly used antisymmetric roundings are rounding to zero, rounding away from zero, or rounding to the nearest floating point number. The approximating operations $\boxplus, \boxminus, \boxdot$, and $\boxslash$ for floating point numbers are required to satisfy

(RG) $a \boxcircle b = \Box(a \circ b)$,   for all $a, b \in R$, and $\circ \in \{+, -, \cdot, /\}$.

Here $+, -, \cdot$, and $/$ denote the operations for real numbers.

If a mapping satisfies the properties (R1), (R2), (R3), and (RG), we call it a *semimorphism*.

All operations defined by (RG), (R1), and (R2) are of maximum accuracy in the sense that there is no element of R lying between the exact result $a \circ b$ executed in $I\!R$ and its approximation $a \boxcircle b$ executed in R. To realize this, we assume that α and β are adjacent elements of R with the property

$\alpha \leq a \circ b \leq \beta$.

Applying (R2), (R1), and (RG) we get:

$\alpha \leq a \boxcircle b \leq \beta$.

That is, $a \boxcircle b$ is either equal to α or to β.

For special applications, a programming language should also provide the directed roundings $\triangledown$ and $\triangle$ which are defined by the properties (R1), (R2), and

(R4) $\triangledown a \leq a$ or $a \leq \triangle a$,   for all $a \in I\!R$.

These roundings, as well as the operations defined by (RG)

$$a \mathbin{\underline{\triangledown}} b := \nabla(a \circ b)$$

or

$$a \mathbin{\underline{\triangle}} b := \triangle(a \circ b)$$

are determined unique ([24], [28]).

Besides the real numbers, vectors and matrices over the real numbers frequently appear in scientific computation. We denote these sets by $V\!I\!R$ and $M\!I\!R$, respectively. Occasionally, the complex numbers $\mathbb{C}$, vectors $V\mathbb{C}$ and matrices $M\mathbb{C}$ over the complex numbers are used. All these spaces are ordered according to the order relation $\leq$, which is defined componentwise in the product spaces. Intervals may be defined via this relation. Numerical algorithms often use intervals in the above-mentioned spaces. If we denote the set of intervals over an ordered set by a preceding I, the spaces $I\!I\!R, IV\!I\!R, I M\!I\!R$, and $I\!\mathbb{C}, IV\mathbb{C}$, and $I M\mathbb{C}$ occur. All these spaces are listed in the first column of the following table. Their subsets, which may be represented on a computer, are described by the symbols listed in the second column of the following table.

Table 1:   Vector Spaces for Scientific Computation

Basic Spaces of Scientific Computation	Subsets Representable on the Computer
$I\!R$	R
$V\!I\!R$	VR
$M\!I\!R$	MR
$I\!I\!R$	IR
$IV\!I\!R$	VIR
$I M\!I\!R$	MIR
$\mathbb{C}$	CR
$V\mathbb{C}$	VCR
$M\mathbb{C}$	MCR
$I\!\mathbb{C}$	CIR
$IV\mathbb{C}$	$VCIR$
$I M\mathbb{C}$	$MCIR$

Next, we define the arithmetic for a vector processor (in our mathematical sense) for all inner and outer operations occurring in the second column of Table 1. We demand that all these operations fulfill the properties of semimorphism. Since these operations in the product spaces differ essentially from the operations executed traditionally on a computer, we would like to briefly repeat their definition here.

Let S be an element of the left column of Table 1, and let T be the corresponding subset in the right column. Furthermore, let a rounding

$$\square : S \to T$$

be given that rounds the elements of S into those of T. Again, this rounding is assumed to satisfy

$\quad$ (R1)$\quad$ $\square a = a,$ $\quad$ for all $a \in T,$ and $\hspace{4cm}$ (projection)

$\quad$ (R2)$\quad$ $a \le b \Rightarrow \square a \le \square b,$ $\quad$ for all $a, b \in S.$ $\hspace{2.5cm}$ (monotonicity)

This rounding is called antisymmetric, if it also satisfies

$\quad$ (R3)$\quad$ $\square(-a) = -\square a,$ $\quad$ for all $a \in S.$ $\hspace{3.5cm}$ (antisymmetry)

The operations in T are defined by

$\quad$ (RG)$\quad$ $a \boxdot b := \square(a \circ b),$ $\quad$ for all $a, b \in T,$ and $\circ \in \{+, -, \cdot, /\},$

with $\circ$ denoting the exact operations in S in the mathematical sense, if they exist.

$\quad$ The operations defined in the product spaces (e.g. for complex matrices) are again of maximum accuracy in the sense that there is no element of T lying between the exact result $a \circ b$ executed in S and its approximation $a \boxdot b$ in T.

$\quad$ In case of the interval spaces occurring in Table 1, the order relation $\le$ in (R2) is replaced by the inclusion relation $\subseteq$. The rounding $\square : IS \to IT$ is assumed to satisfy the additional property

$\quad$ (R4)$\quad$ $a \subseteq \square a,$ $\quad$ for all $a \in IS.$ $\hspace{3.5cm}$ (upwardly directed)

The theory developed in [24] and [28] shows that this rounding is uniquely defined.

$\quad$ The usual definition of computer arithmetic differs considerably from our definition. Traditionally, computer arithmetic comprises only the operations in R. All other operations in the second column of Table 1 have to be implemented by the user. In general, this is done by procedures where every operation occurring in an algorithm requires its own procedure call. This procedure is cumbersome, inefficient for both programmers and computers, and often subject to inaccuracies. Let us consider the example of the matrix multiplication $C = A \cdot B$, requiring the execution of a scalar product for each component of C. This is usually done on the basis of real floating point operations of the form

$$C = (c_{ij}) = (a_{i1} \boxdot b_{1j} \boxplus a_{i2} \boxdot b_{2j} \boxplus \dots \boxplus a_{in} \boxdot b_{nj})$$

with $2n - 1$ roundings. By contrast, the formula (RG) requires an implementation of the rule $C = A \square B$ with

$$C = (c_{ij}) = (\square(a_{i1} \cdot b_{1j} + a_{i2} \cdot b_{2j} + \dots + a_{in} \cdot b_{nj}))$$

with only one rounding for each component. A computer which satisfies our axioms calculates this formula for arbitrary n with one single rounding. This *optimal scalar product* plays a significant part in all product spaces of Table 1. In addition, the scalar product is very often used in numerical algorithms to achieve high accuracy.

A vector processor in the mathematical sense, as we define it here, is required to provide all inner and outer operations in the spaces of the right column of Table 1 by semimorphism. In PASCAL–XSC, the sets in the right column of Tabel 1 are predefined types. Variables and values of these types may be combined by means of the usual operator symbols +, -, *, and /. In PASCAL–XSC, the operator symbols denote the operations ⊞, ⊟, ☐, ⍁ used above and defined by means of semimorphism. Expressions of these types may be written down easily and clearly. All inner and outer operations fulfill the demands of the semimorphism. The arithmetic we are describing can be implemented in hardware, in firmware, or in software. If the basic hardware of the processor in use supports these requirements, the operations are even faster than those traditionally implemented and executed via operations in R. In case an appropriate support by the hardware is lacking, the operations according to semimorphism in the spaces of the second column of Table 1 are simulated in software via the run-time system of PASCAL–XSC. The software arithmetic of PASCAL–XSC is realized using *integer* arithmetic. In case an IEEE arithmetic coprocessor is available, its operations can be used. However, a software simulation of the optimal dot product for accumulation of numbers and products is still necessary. The latter is the most useful operation for automatic verification of computed results.

Of course, a software simulation of the arithmetic increases the execution time. On the other hand, the user has a well defined and comprehensive arithmetic. He can fully rely on its properties and build upon them in numerical algorithms. From the perspective of arithmetic and programming languages, PASCAL–XSC is an ideal vector language. Programming of algorithms in engineering and scientific computation is facilitated by the language extensions. PASCAL–XSC is particularly suited for the development of numerical algorithms that deliver highly accurate and automatically verified results.

1.4   Connection with Programming Languages

The demands of a high quality computer arithmetic lead quite naturally to the concepts of a programming language for vector processors like PASCAL–XSC or ACRITH–XSC. Usual programming languages like ALGOL, FORTRAN, PASCAL, MODULA, or PL/1 possess only the *integer*, *real*, and (perhaps) *complex* arithmetic as elementary operations. All other arithmetic operations, especially those in the product spaces shown in the second column of Table 1, have to be based upon the *integer* and *real* arithmetic.

In contrast, PASCAL–XSC provides all operations in the product spaces for predefined types via the usual operator symbols. Each of these operations calls elementary operations which are implemented directly and with maximum accuracy.

In general, the operations in the product spaces could even be carried out in parallel. Unlike the other languages listed above, PASCAL–XSC provides the following language elements and features:

- Explicit language support for the directed roundings $\triangledown$ and $\triangle$.
- Explicit language support for the corresponding operations $\underline{\triangledown}$ and $\underline{\triangle}$ for all $\circ \in \{+, -, \cdot, /\}$.
- An optimal scalar product for vectors of arbitrary length.
- Interval types with appropriate operators.
- Functions with arbitrary result type.
- A universal operator concept.
- Overloading of function identifiers and operators.
- Dynamic and structured numerical types
- A large number of mathematical functions with high accuracy for the numerical types *real, complex, interval*, and *complex interval*.

A library of problem solving routines with results of highest accuracy and automatic result verification (see also [27]) for many standard problems of numerical mathematics has been implemented in PASCAL–XSC. Via interval input data the accuracy of the elementary functions becomes immediately visible to the user.

PASCAL–XSC is a true vector language in the mathematical sense. The vector notation of the operations in the product spaces is already expressed in the programming language. An additional vectorization of programs by the compiler is often superfluous. The execution of these operations may be essentially accelerated by parallel processing and pipeline techniques.

1.5   Survey of PASCAL–XSC

The programming language PASCAL–XSC was developed to supply a powerful tool for the numerical solution of scientific problems based upon a properly defined and implemented computer arithmetic in the usual spaces of numerical computation (see [24], [28]). The main concepts of PASCAL–XSC are

- ISO Standard PASCAL
- Universal operator concept (user-defined operators)
- Functions and operators with arbitrary result type
- Overloading of procedures, functions, and operators
- Module concept
- Dynamic arrays
- Access to subarrays

- String concept
- Controlled rounding
- Optimal (exact) scalar product
- Predefined type *dotprecision* (a fixed point format to cover the whole range of floating-point products)
- Additional arithmetic built-in types such as *complex, interval, rvector, rmatrix*, etc.
- Highly accurate arithmetic for all built-in types
- Highly accurate elementary functions
- Exact evaluation of expressions within accurate expressions (#-expressions)

Interval arithmetic, complex arithmetic, complex interval arithmetic, and the corresponding vector and matrix arithmetics are provided.

Application modules have been implemented in PASCAL–XSC for solving common numerical problems, such as

- Linear systems of equations
- Matrix inversion
- Nonlinear systems of equations
- Eigenvalues and eigenvectors
- Evaluation of arithmetic expressions
- Evaluation of polynomials and zeros of polynomials
- Numerical quadrature
- Initial and boundary value problems in ordinary differential equations
- Integral equations
- Automatic differentiation
- Optimization problems

All these problem-solving routines provide *automatically verified results*.

In the subsequent sections, the most important new concepts are considered briefly. The details are described in chapter 2.

1.5.1 Universal Operator Concept and Arbitrary Result Type

PASCAL–XSC makes programming easier by allowing the programmer to define functions and operators with arbitrary result type. The advantages of these concepts are illustrated by the simple example of polynomial addition. Define the type `polynomial` by

```
const maximum_degree = 20;
type  polynomial = array [0..maximum_degree] of real;
```

in Standard PASCAL, the addition of two polynomials is implemented as a *procedure*

```
procedure add (a, b: polynomial; var c: polynomial);
  { Computes c = a + b for polynomials }
  var i: integer;
  begin
    for i:= 0 to maximum_degree do
      c[i]:= a[i] + b[i];
  end;
```

Several calls of add have to be used to compute the expression $z = a + b + c + d$:

```
add (a,b,z);
add (z,c,z);
add (z,d,z);
```

In PASCAL–XSC, we define a *function* with the result type *polynomial*

```
function add (a, b: polynomial): polynomial;
  { Delivers the sum a + b for polynomials }
  var i: integer;
  begin
    for i:= 0 to maximum_degree do
      add[i]:= a[i] + b[i];
  end;
```

Now, the expression $z = a + b + c + d$ may be computed as

```
z:= add(a,add(b,add(c,d))).
```

Even clearer is the *operator* in PASCAL–XSC

```
operator + (a, b: polynomial) result_polynomial : polynomial;
  { Delivers the sum a + b for polynomials }
  var i: integer;
  begin
    for i:= 0 to maximum_degree do
      result_polynomial[i]:= a[i] + b[i];
  end;
```

Now, the expression may be written in the common mathematical notation

```
z:= a+b+c+d.
```

A programmer may also define a new name as an operator. A priority is assigned in a preceding priority declaration.

1.5.2   Overloading of Procedures, Functions, and Operators

PASCAL–XSC permits the overloading of function and procedure identifiers. A generic name concept allows the programmer to apply the identifiers *sin, cos, exp, ln, arctan*, and *sqrt* not only for *real* numbers but also for intervals, complex numbers, or elements of other mathematical spaces. Overloaded functions and procedures are distinguished by number, order, and type of their parameters. The result type is *not* used for distinction.

As illustrated above, operators may also be overloaded. Even the assignment operator (:=) may be overloaded so that the mathematical notation may be used for assignments:

```
var
  c: complex;
  r: real;

operator := (var c : complex; r: real);
  begin
    c.re := r;
    c.im := 0;
  end;

...

r:= 1.5;
c:= r;  {complex number with real part 1.5 and imaginary part 0}
```

1.5.3   Module Concept

The module concept allows the programmer to separate large programs into modules and to develop and compile them independently of each other. The control of syntax and semantics may be carried out beyond the bounds of the modules. Modules are introduced by the reserved word **module** followed by a name and a semicolon. The body of a module is built up quite similarly to that of a common PASCAL program. The significant exception is that the objects to be exported from the module are characterized by the reserved word **global** directly in front of the reserved words **const, type, var, procedure, function**, and **operator** and directly after **use** and the equality sign in type declarations. Thus, private types as well as non-private types can be declared and exported.

Modules are *imported* into other modules or programs via a **use**-clause. The semantics of the **use**-clause are that all objects declared **global** in the imported module are also known in the importing module or program.

The example of a polynomial arithmetic module illustrates the structure of a module:

```
module poly;
  use {other modules}
  ...
  {local declarations}
  ...
  {global declarations}
  global type  polynomial = ...
  ...
  global procedure read (...
  ...
  global procedure write (...
  ...
  global operator + (...
  ...
  global operator * (...
  ...
begin
  {initialization part of the module}
  ...
end. {module poly}
```

1.5.4 Dynamic Arrays and Subarrays

The concept of dynamic arrays enables the programmer to implement algorithms independently of the length of the arrays used. The index ranges of dynamic arrays are not to be defined until run-time. Procedures, functions, and operators may be programmed in a fully dynamic manner, since allocation and release of local dynamic variables are executed automatically. Hence, the memory is used optimally.

For example, a dynamic type polynomial may be declared in the following form:

```
type  polynomial = dynamic array [*] of real;
```

When declaring variables of this dynamic type, the index bounds have to be specified:

```
var  p, q : polynomial [0..2*n];
```

where the values of the expressions for the index range are computed during program execution. To get access to the bounds of dynamic arrays which are specified only during execution of the program, the two functions *lbound(...)* and *ubound(...)* and their abbreviations *lb(...)* and *ub(...)* are supplied. The multiplication of two polynomials may be realized dynamically as follows:

```
operator * (a, b: polynomial)
          product: polynomial[0..ubound(a)+ubound(b)];
  { Delivers the product a * b of two polynomials a, b }
  var i, j   : integer;
      result : polynomial[0..ubound(a)+ubound(b)];
```

```
    begin
      for i:= 0 to ubound(a)+ubound(b) do
        result[i]:= 0;
      for i:= 0 to ubound(a) do
        for j:= 0 to ubound(b) do
          result[i+j]:= result[i+j] + a[i] * b[j];
      product:= result;
    end;
```

A PASCAL–XSC program using dynamic arrays for polynomials follows the template

```
  program dynatest (input, output);
  ...
  type polynomial = dynamic array [*] of real;
  ...
  var  maximum_degree : integer;
  ...
  operator * (a, b:polynomial)...
  ...
  procedure write (var f : text; p: polynomial);
  ...
  procedure main (degree : integer);
    var
      p,q : polynomial[0..degree];
      r   : polynomial[0..2*degree];
    begin
      ...
      r:= p * q;
      writeln ('p*q = ', r);
      ...
    end;

  begin {main program}
    read (maximum_degree);
    main (maximum_degree);
  end. {main program}
```

The following example demonstrates that it is possible to access a row or a column of dynamic arrays as a single object. This is called *slice* notation.

```
  type vector =  dynamic array [*] of real;
  type matrix =  dynamic array [*] of vector;
  var  v : vector[1..n];
       m : matrix[1..n,1..n];
  ...
  v        := m[i];   { i-th row of m    }
  m[*,j] := v;        { j-th column of m }
```

1.5.5   String Concept

A string concept is integrated into the language PASCAL–XSC to handle character strings of variable length. Declaration, input, and output of strings are simplified extensively compared to the facilities of Standard PASCAL. Special predefined functions and operators enable the programmer to formulate and manipulate string expressions.

1.5.6   Arithmetic and Rounding

Compared with Standard PASCAL, the set of operators for *real* numbers is extended by the directed-rounding operators $\circ<$ and $\circ>$ with $\circ \in \{+, -, *, /\}$. These symbols denote the operations with upwardly and downwardly directed rounding. In the arithmetic modules, the common operators are also provided for complex numbers, intervals, complex intervals, and also for vectors and matrices over these spaces.

1.5.7   Accurate Expressions

The implementation of inclusion algorithms with high accuracy requires the exact evaluation of scalar products. For this purpose, the new type *dotprecision* was introduced into PASCAL–XSC representing a fixed-point format covering the whole range of floating-point products. This format allows scalar results – especially sums of floating-point products – to be stored exactly.

Furthermore, scalar product expressions (dot product expressions) with vector and matrix operands with only one rounding per component can be computed via exact evaluation within accurate expressions (#-expressions).

Chapter 2

Language Reference

PASCAL–XSC is based on the programming language PASCAL defined in the report of Jensen and Wirth [13]. Since PASCAL–XSC is an extension of PASCAL, we do not give a detailed description of the complete language (for this purpose see [9], [13], or [14] for example). Instead, we give a concise description of the standard elements of PASCAL and a rather more detailed introduction into the additional language elements of PASCAL–XSC.

The syntax is specified in an easy readable, simplified Backus-Naur-Form, very similar to usual program code. It is marked by a black bar at the margin. The representation of the syntax consists of

- basic symbols written according to the typographical notation of section 2.1,
- syntax variables,
- predefined identifiers,
- repetition symbols ... , and
- comments enclosed in braces { }.

Syntax variables are English nouns or composite nouns which serve as abbreviations for other syntactical units. If a syntax variable denotes a list, then it stands for a non-empty sequence of corresponding objects separated by commas. Predefined identifiers are written in slanted characters. The repetition symbol ... denotes an arbitrary number of repetitions of a part of the syntax, i.e. the preceding group of language elements within the same line. This group may occur zero or more times if no comment to the contrary (enclosed in braces) is given.

To generate PASCAL code by means of this syntax representation, we have to read line after line from left to right and note the basic symbols, the predefined identifiers, and the syntax variables in the suitable number of repetitions. Then to eliminate the syntax variables, we must successively replace them by their own definitions. This process ends when all syntax variables are eliminated.

An example for a syntax description is

```
var
    IdentifierList: Type; ... { not empty }
```

Assuming that we know what to replace for Type and what kind of Identifiers are allowed, we consider an example:

Example 2.0.1:

> The syntax for variable declarations in the example above allows the following source code:

var i, j, k : integer;
    x, y : real;
    m   : **array** [1..10, 1..10] **of** real;

The extensions provided by PASCAL–XSC beyond ISO Standard PASCAL are highlighted in this chapter by a box:

```
────────────────────────────────────── PASCAL–XSC ──

All extensions provided by PASCAL–XSC beyond the ISO Standard PASCAL
are enclosed in frames like this!
```

A concise survey of the complete syntax of the language PASCAL–XSC is given by the *syntax diagrams* in Appendix A starting from 269.

For all implementation-dependencies mentioned in this chapter and in the following chapters, we refer to the corresponding user manual supplied with the special compiler version.

2.1 Basic Symbols

The source code of a program consists of the following basic symbols:

letters: a, b, c, ..., z

digits: 0, 1, 2, ..., 9

special characters: `<=   <   >   >=   =   <>`
                      `(   )   [   ]   {   }`
                      `+   −   *   /`
                      `:=   .   ,   ;   :   '   ↑   ⊔   ..`

The character ⊔ denotes a blank.

Instead of   `{   }   [   ]   ↑`
the characters `(*  *)  (.  .)  @ or ^` may be used.

reserved words: **and, array, begin, case, const, div, do, downto, else, end, file, for, forward, function, goto, if, in, label, mod, nil, not, of, or, packed, procedure, program, record, repeat, set, then, to, type, until, var, while, with**

Letters may be used in either upper or lower case, but they are treated as equivalent. For example, the identifiers *PASCAL* and *Pascal* are identical. A reserved word may be written in any mixture of upper case or lower case letters. PASCAL–XSC is *not* case sensitive as C is.

Reserved words may not be used as identifiers.

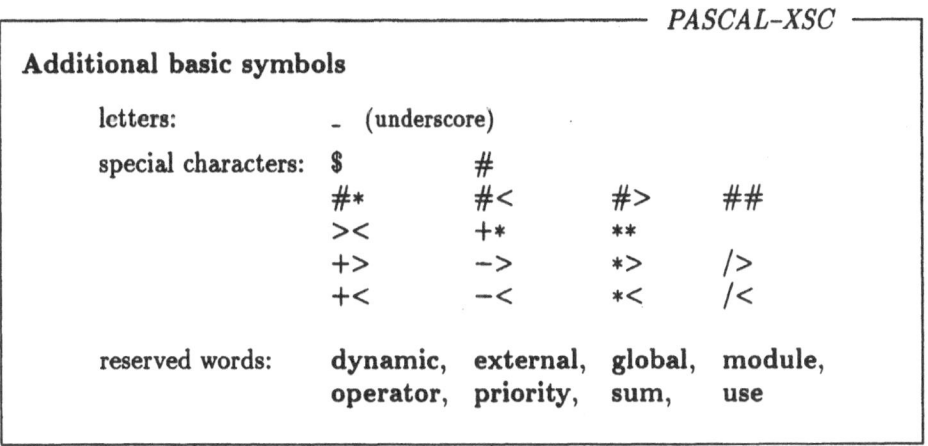

PASCAL–XSC

Additional basic symbols

letters: _ (underscore)

special characters: `$        #`
                      `#*       #<       #>       ##`
                      `><       +*       **`
                      `+>       −>       *>       />`
                      `+<       −<       *<       /<`

reserved words: **dynamic, external, global, module, operator, priority, sum, use**

2.2   Identifiers

Identifiers are names denoting different objects like constants, variables, types, functions, procedures, modules etc. occurring within a program.

An identifier consists of an arbitrary sequence of letters and digits beginning with a letter.

Example 2.2.1:

variable1, NorthWest, extreme, ab

Two identifiers are considered to be identical if they both consist of the same sequence of characters, ignoring the case of letters.

Reserved words may *not* be used as identifiers. When defining a new identifier, note that the following identifiers have predefined meanings:

abs	eof	ln	pred	round	trunc
arctan	eoln	maxint	put	sin	write
boolean	exp	new	read	sqr	writeln
char	false	odd	readln	sqrt	
chr	get	ord	real	succ	
cos	input	output	reset	text	
dispose	integer	page	rewrite	true	

These identifiers can be used in their predefined meaning without being declared explicitly. If they are given a new meaning by an explicit declaration, any occurrence of such an identifier refers to the new meaning. The usual meaning provided by the language becomes invisible.

PASCAL–XSC

The underscore (_) may occur at any position of an identifier.

Example 2.2.2:

variable_1, north_west, _extreme_, a__b, _, __

The maximum length of an identifier is the logical line length.

Upper case and lower case letters are not distinguished within identifiers. This means that *north_west* and *North_West* denote the same identifier.

PASCAL–XSC

Additional predefined identifiers:

arccos	cmatrix	im	log2	setlength
arccot	comp	image	log10	sign
arcsin	complex	imatrix	mant	sinh
arcosh	cosh	inf	mark	string
arcoth	cot	interval	maxlength	substring
arctan2	coth	ivector	pos	sup
arsinh	cvector	ival	re	tan
artanh	dotprecision	lb	release	tanh
cimatrix	expo	lbound	rmatrix	ub
cinterval	exp2	length	rvector	ubound
civector	exp10	loc	rval	

The predefined identifiers of procedures and functions may be overloaded. By this device, they may be used in the predefined meaning as well as in the new meaning (see section 2.7.10). However, if they are declared in a manner identical to the predefined declaration, then the predefined meaning becomes invisible, and the new meaning is used.

When using a module by means of a **use**-clause, all identifiers declared in the used module via **global** become predefined identifiers in the using module or program (see section 2.8).

Identifiers are also used to denote operators (see section 2.7.6).

2.3   Constants, Types, and Variables

In ISO Standard PASCAL, the types *integer, real, boolean*, and *char* are predefined together with corresponding operators (see section 2.4). In addition, an enumeration type may be declared which defines a set of entirely new values and introduces constant identifiers to denote these values. The values of such simple types are called (literal) constants. Their syntax is precisely defined (see section 2.3.1). Except for string constants (in the case of a one dimensional array with *char* as component type), there are no literal constants for the structured types array, set, record, and file. A literal constant of a string is a sequence of characters enclosed in single quotes.

Identifiers for constants (named constants) may be introduced by a constant definition:

> **const**
>   Identifier = Constant; ... { not empty }

The constant on the right-hand side must be a literal constant or a previously defined named constant.

Example 2.3.1:

```
const
    n      = 50;
    eps    = 10e−13;
    k      = n;
    zf     = 'charactersequence';
```

Named constants may be used like literal constants within a program. During execution of a program, they are unchangeable.

Identifiers for types (named types) may be fixed by a type definition:

> **type**
>   Identifier = Type; ... { not empty }

The type on the right-hand side is either an explicitly defined type or a previously defined type. All types mentioned above are always predefined.

Example 2.3.2:

```
type  color    =   (red, blue, yellow);
      logical  =   boolean;
      vector   =   array [1..20] of real;
```

Identifiers for variables may be fixed by a variable declaration :

> **var**
>   IdentifierList: Type; ... { not empty}

The listed identifiers denote variables with the type of the right-hand side. A variable may therefore be interpreted as a symbolic address of a corresponding storage space.

Example 2.3.3:

```
var      i, j, k  :  integer;
          x, y  :  real;
             f  :  color;
    vec1, vec2  :  vector;
             m  :  array [1..20] of vector;
```

```
──────────────────────────────── PASCAL–XSC ───
```

The following additional standard types are available: *dotprecision, complex, interval, cinterval, rvector, cvector, ivector, civector, rmatrix, cmatrix, imatrix, cimatrix,* and *string*.

  A dynamic array type within a variable declaration must specify the index bounds by corresponding expressions (see section 2.3.2 concerning dynamic array types).

  A variable is called of *anonymous* type if there is no related type identifier in the corresponding declaration (e.g. component variable).

Example 2.3.4:

```
var  vec1, vec2  :  vector;            { known type }
              a  :  array [1..10] of real;   { anonymous type }
              m  :  array [1..20] of vector;  { anonymous type }
```

2.3.1 Simple Types

The simple types in PASCAL are the types *integer, real, boolean, char,* enumeration types, and subrange types. They are defined as follows:

integer   Implementation-dependent subset of the whole numbers. The predefined constant *maxint* denotes the implementation-dependent maximum integer number. A literal constant of type *integer* is a digit sequence (of decimal digits) with or without a sign + or −. See page 23 for PASCAL–XSC extensions of type *integer*.

Example 2.3.5:

128   −30   +4728   007

real

Implementation-dependent subset R of the real numbers $\mathbb{R}$. A literal constant of type *real* has the representation

$$\pm \text{ mantissa E exponent}$$

The mantissa is a sequence of digits with or without a decimal point, and the exponent is an *integer* value (with implementation depended bounds). The notation

$$\pm \text{ mantissa}$$

without an exponent part is permitted as well. As a matter of principle, at least one digit must occur in front of and behind the decimal point.

Example 2.3.6:

3.1726E−2   −0.08E+5   +1E10   3.1415

We have to bear in mind that the value of the decimal representation in PASCAL is not always a member of the implementation-dependent set R of type *real*. For example, the decimal floating-point number 1.1 is not exactly representable as binary floating-point number. Thus, the literal constant 1.1 used within a program represents a *real* value different from the value 1.1. This problematic nature of conversion we have to bear in mind every time we use literal constants as operands within expressions, as parameters in function and procedure calls, or as input parameters. See page 24 for the PASCAL–XSC extensions of type *real*.

boolean

The range consists of the logical constants *true* and *false* with *false* < *true*.

char

The range is an implementation-dependent set of characters. Literal constants are enclosed in single quotes. The order relation satisfies

$$'0' < '1' < \ldots < '9'$$

and

$$'a' < 'b' < \ldots < 'z'.$$

enumeration types   The range consists of constants listed in the type definition (ordered sequence of identifiers). The order relation is defined by the order of the enumeration. An enumeration type is defined by the programmer in a type definition. The literal constants which are the elements of an enumeration type must not collide with any value of another enumeration type.

Example 2.3.7:

The type definition

type Color = (red, blue, yellow);

specifies the enumeration type *Color* with the values *red, blue, yellow*. Another type

SpecialColor = (yellow, orange);

is not permitted since the value *yellow* already occurs in type *Color*.

subrange types   Subrange types of each of the predefined types *integer, boolean, char*, and all enumeration types (base types) may be defined by specifying the lower and upper bound by

constant .. constant

The set of values of a subrange consists of the lower and upper bound and all values of the base type between them. The important thing here is that the order relation is inherited from the base type. The lower bound must be less than or equal to the upper bound, and both must be of the same base type.

Example 2.3.8:

```
type
  Subrange    =       1..100;
  SubColor    =   blue..yellow;
  Letters     =     'a'..'z';
  OctalDigits =       0..7;
```

                                                          PASCAL–XSC

PASCAL–XSC allows some additional notations for the types *integer* and *real*. Furthermore, we introduce the new type *dotprecision*.

integer   A value of type *integer* may also be written as a hexadecimal constant beginning with the character $ and followed by a hexadecimal digit sequence consisting of the digits $0, 1, \ldots, 9$ and the letters A,B,...,F and a,b,...,f.

────────────────────────────────── *PASCAL–XSC* ──────

Example 2.3.9:

$12AFB2   represents the value   1224626

real   To execute the inevitable conversion of literal *real* constants into the internal data format in a controlled way, an additional notation for these constants is necessary. While the usual PASCAL notation of *real* numbers implies the conversion with rounding to the nearest floating-point (machine) number, it is possible to specify *real* constants which are converted with rounding to the next-smaller or the next-larger floating-point number by the notations

$$(< \pm \text{ Mantissa E Exponent })\qquad\text{and}$$
$$(> \pm \text{ Mantissa E Exponent }),$$

respectively. The E and Exponent may be omitted as usual, in which case Mantissa must contain a decimal point. The parentheses are mandatory.

Example 2.3.10:

$$(< 1.1)\qquad\text{round down}$$
$$(> -1.0\text{E}-1)\qquad\text{round up}$$

dotprecision   The type *dotprecision* is based on the type *real* and permits the representation of products of two arbitrary real numbers and the exact summation of an *arbitrary* number of such products in a fixed point format of suitable size. If the internal *real* format is fixed by the mantissa length l and the minimum and maximal exponents *emin* and *emax* (see also section 2.4.1.2), then a *dotprecision* variable occupies storage space of the form

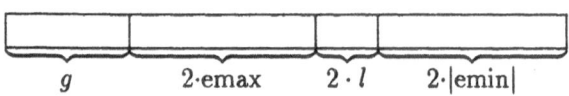

$$g \qquad 2\cdot emax \qquad 2\cdot l \qquad 2\cdot|emin|$$

The total length is $L = g + 2emax + 2|emin| + 2l$ digits. Here, g denotes the implementation-dependent number of guard digits for accumulating carries during summation (see [28] and [29] for details).

────────────────────────────── *PASCAL–XSC* ──────

Values of type *dotprecision* typically occur during multiplication
of vectors or matrices. These scalar products can be represented
exactly, i.e. without rounding errors, in the format of this type independent of the dimensions of the vectors or matrices. A Value
of type *dotprecision* can only be generated by an #-expression
(see section 2.4.2) in the form

(ExactExpression)

There are no constants of this type.

2.3.2 Structured Types

There are four composite data types in ISO Standard PASCAL:

- arrays

- files

- records

- sets

They differ in their manner of combining elements of the predefined types into
a higher structure and accessing their components. Composite types of arbitrary
complexity may be built up by using components of arbitrary composite types.

Any type definition may start with the reserved word **packed** which causes
the components to be stored in a compact fashion. The storage pattern is
implementation-dependent. The reserved word **packed** has no semantic effect.

2.3.2.1 Arrays

An array consists of a fixed number of components having the same type. Each
component is indicated by one or more indices which are values of index expressions.
The type definition of an array must specify index types and the component type:

array [IndexTypeList] **of** ComponentType

An index type is a subrange type of type *integer*, *boolean*, *char*, or an enumeration
type, or of one of the last three types itself. The component type may be of any
type. Note that the use of the component type *dotprecision* may be very memory-
consuming.

Example 2.3.11:

array [1..10] **of** real	{ real vector with }
	{ 10 components }
array [1..10, 1..20] **of** real	{ real matrix with }
	{ 10 rows and 20 columns }
array ['a'..'z'] **of** boolean	{ logical vector with }
	{ 26 components }
array [1..10] **of array** ['a'..'z'] **of** boolean	{ logical matrix with }
	{ 10 rows and 26 columns }
array [(red, yellow, blue, black)] **of** 0..10	{ subrange type vector with }
	{ 4 components }

The components of an array themselves may be used as variables (component variables). The access to these component variables is done by

ArrayIdentifer [IndexExpressionList]

or

ArrayIdentifier[IndexExpressionList]
                        [IndexExpressionList] ...

The relation of index expressions (indices) and index ranges works from left to right. The indices must be contained in the corresponding index range.

Example 2.3.12:

declaration	component variables
var v: **array** [1..10] **of** real;	v[1], ..., v[10]
var m: **array** [1..10,1..20] **of** real;	m[1,1], ..., m[1,20],
	m[2,1], ..., m[2,20],
	$\vdots$
	m[10,1], ..., m[10,20]
	also possible:
	m[1][1], ...
var x: **array** [1..10] **of** **array** ['a'..'z'] **of** boolean;	x[1]['a'], ..., x[1]['z'],
	$\vdots$
	x[10]['a'],..., x[10]['z']
	also possible
	x[1,'a'], ...
	x[1,a] only works if a is a variable of type *char*.

2.3.2.2 Subarrays

If the number k of specified indices of a component variable of an array is less than the number n of index ranges of the corresponding array (n dimensional array), then the component variable is an $n - k$ dimensional subarray. The specified indices refer to the first k index ranges within the array declaration.

--- *PASCAL–XSC* ---

A component variable is called of *anonymous type* if there is no corresponding explicit type identifier (see also section 2.3.5 referring to compatibility of types).

Example 2.3.13:

> Let the variable m be a two-dimensional array declared by
>
> **var** m: **array** $[1..10, 1..20]$ **of** real;
>
> Then the component variable $m[5]$ is a one-dimensional subarray of anonymous type, i.e. a vector of 20 components consisting of the row of the matrix m.

Arbitrary subarrays (component variables) of an array may be accessed by specifying corresponding index ranges by placing the character * within the index expression list. If there is no index expression following a *, then it may be omitted.

Example 2.3.14:

> According to the declaration
>
> **var** m: **array** $[1..10, 1..20]$ **of** real;
>
> the component variable $m[*,1]$ denotes an array variable, i.e. a vector with 10 elements, where the elements correspond to the elements of the first column of the matrix m.
>
> The notations
>
> $m[1,*]$ and $m[1]$ as well as
>
> m   and $m[*]$ and $m[*,*]$
>
> are equivalent.

2.3.2.3   Access to Index Bounds

––– *PASCAL–XSC* –––––

The access to index bounds without using the quantities of the array declaration (absolutely necessary in case of using dynamic arrays) is provided by two standard functions

> lbound (ArrayVariable, IntegerConstant) and
> ubound (ArrayVariable, IntegerConstant).

Their result is the lower and upper bound of the i-th index range (i = value of the integer constant) of the array variable. In case the integer constant is missing, the first index range is addressed implicitly. The access to index ranges that do not exist is not allowed.

The notations *lb* (instead of *lbound*) and *ub* (instead of *ubound*) may be used as short forms.

Example 2.3.15:

```
type matrix = array [1..n,1..k] of real;
function sum(var m: matrix): real;
var
    i, j: integer;
    s: real;
begin
    s := 0;
    for i := lbound(m) to ubound(m) do
        for j:= lb(m,2) to ub(m,2) do
            s := s + m[i,j];
    sum:= s;
end;
```

2.3.2.4   Dynamic Arrays

ISO Standard PASCAL does not provide dynamic arrays. A certain dynamic capability is given in level 1 of the standard (see [9]) by *conformant array schemes* as specification of array parameters within functions and procedures (see section 2.7.1). Thus, the call of the corresponding procedures and functions is possible with actual parameters which need not necessarily be of a definite type.

––– *PASCAL–XSC* –––––

PASCAL–XSC provides the *dynamic array declaration*, similar to that provided by other well-known programming languages (e.g. ALGOL 60, ALGOL 68 [49], ADA [15]). This means that within subroutines, array variables need not be declared statically as in Standard PASCAL.

—————————————————————————— *PASCAL–XSC* ———

The index bounds may be given by expressions which result in new index bounds at each subroutine call. Especially, the dynamic array concept and its use in the specification of formal parameters includes the complete functionality of the conformant array scheme of Standard PASCAL using only a slightly different syntactical notation.

A dynamic array type may be declared similar to the declaration of a static array type:

| **dynamic array** [DimensionList] **of** ComponentType

Every index range within the dimension list is marked by the character ∗.

A dynamic array type must not occur as component type of a structured type, except in a dynamic array type itself. The dynamic array type definition specifies only the number of the index ranges and the component type.

Example 2.3.16:

> **type**   DynPolynomial   =   **dynamic array** [∗] **of** real;
> DynVector   =   **dynamic array** [∗] **of** real;
> DynMatrix   =   **dynamic array** [∗,∗] **of** real;

> Not allowed is

> **type**   WrongType   =   **dynamic array** [1..n,∗] **of** real;

Dynamic array types may occur within a variable declaration by specifying corresponding index expressions either by using a previously declared type identifier or by explicitly denoting the dynamic type.

Example 2.3.17:

> **var**   mat1:   dynmatrix [1..n,1..2∗n];
> mat2:   **dynamic array** [1..n,1..2∗n] **of** real;

In both cases, the computation of the corresponding index expressions must be determined at the time of processing of the variable declaration. Thus, true dynamic allocation in array declaration is only possible within procedures and functions by using global quantities or formal parameters in the index expressions (see section 2.10).

The declaration part of a program can contain expressions in the index bounds only if they can be evaluated at time of declaration. The reserved word **packed** must not occur in a dynamic array declaration, i.e. sequences like **packed dynamic array** or **dynamic packed array** are not permitted.

2.3.2.5 Strings

The string type of ISO Standard PASCAL is a special static packed array type, a
vector with the component type char:

| **packed array** [1..Length] **of** char

with the *integer* constant *Length* ≥ 1.

Example 2.3.18:

The declaration

packed array [1..15] of char

defines strings with 15 characters of type *char*. Examples for constants of this
type are

'PASCAL–XSC TOPS'   or   'string constant'

with exactly 15 characters occurring between the two quotes.

2.3.2.6 Dynamic Strings

PASCAL–XSC

The declaration of string variables is facilitated by the type specification

| string [Length]

or just

| string

Length must be a positive *integer* constant which is bounded by an implementation dependent maximum length, e.g. 255. The maximum length is assumed if
the length specification is missing. The range of this dynamic string type consists
of all character sequences with 0, 1, 2, .., *Length*−1, *Length* characters.

The current length of a string variable is called its actual length. It is dynamically managed at run-time of the program and may be accessed to by the
function *length* and changed by the procedure *setlength* (see section 2.9).

PASCAL-XSC ──

Variables of type _string_ may be indexed. So _s[i]_ denotes the _i_-th character of the string _s_ and is of type _char_. Access to components outside the declared length is not permitted. A change of the actual length of a _string_ variable may only be effected by the procedure _setlength_ or an assignment to the _string_ variable.

See string expressions (section 2.4.3.2) and text processing (section 2.9) for the discussion of the use of dynamic strings.

2.3.2.7   Records

A record is a structure consisting of a fixed number of components, called fields. Fields may be of different types and each field is given a name, the field identifier, which is used to select it. The definition has the form:

> **record** FieldList **end**

A field list is an enumeration of fields of the form

> FieldIdentifierList: Type; ...

Each field identifier in FieldIdentifierList denotes a component of the record. A field identifier is required to be unique only within the record in which it is defined. A field is referenced by the variable identifier and the field identifier separated by a period:

> RecordIdentifier.FieldIdentifier

Example 2.3.19:

```
record
   hour: 0..23;
   minute, second: 0..59;
end;

record
   re, im : real;
end;

var date: record month: (Jan, Feb, Mar, Apr, May, Jun,
                         Jul, Aug, Sep, Oct, Nov, Dec);
              day:    1..31;
              year :  integer;
        end;
```

The component variables of the variable _date_ are accessible by

> date.month   date.day   date.year

2.3.2.8   Records with Variants

A record may be extended by so-called variants, i.e. additional components which are all stored in the same storage space. The programmer has to keep this space under his control. The variants allow the passing of values without the strong type control of PASCAL. In the type definition, the variant part is listed following the fixed components in the form

> **case**
>   TagField: { may be omitted }
>   TagType **of**
>   TagList: (FieldList); ... { not empty }

The tag field is actually a fixed component. It is denoted by an identifier. The tag type is the type of the tag field and of the following tag list elements (constants of the tag type). The types *integer*, *boolean*, *char*, enumeration types, and their subrange types are permitted tag types.

An access to a variant should occur only after *activation* of the desired variant, by the assignment of the corresponding value to the tag field component. If a tag field component is missing, a variant becomes activated by the first access to one of its components.

The components of a variant are accessed like fixed components of a record.

Example 2.3.20:

Let a variant record type *TrafficSignType* be defined by

```
type
    form = (circle, rectangle, triangle);
    TrafficSignType = record
                        serialnumber: integer;
                        material: (metal, synthetic);
                        price: real;
                        case figure: form of
                            circle:    (radius : real);
                            rectangle:(length, height: real);
                            triangle: (baseline, angleleft,
                                        angleright: real);
                    end;
```

A variable *TrafficSign* declared by

```
    var TrafficSign : TrafficSignType;
```

has three variants, the components of which may be accessed by

```
    TrafficSign.figure   :=   circle;
    TrafficSign.radius   :=   3.5;
```

or

  TrafficSign.figure := rectangle;
  TrafficSign.length := 7.8;
  TrafficSign.height := 4.4;

or

  TrafficSign.figure   := triangle;
  TrafficSign.baseline := 5;
  TrafficSign.angleleft := 18.1;
  TrafficSign.angleright := 45;

PASCAL–XSC

With the exception of the type *string*, a dynamic array type may not occur as a component of a record.

2.3.2.9 Sets

The range of values of a set type consists of all subsets of a given basetype. Therefore, the type definition of a set should only specify the base type:

 set of BaseType

A base type may be a subrange of *integer*, *boolean*, *char*, and enumeration type or one of the last three types itself. In most implementations, the ordinal values of the base type must be within the range 0 through 255 ($0 \leq \text{ord}(x) \leq 255$). An access to elements of a set M corresponding to component variables of arrays and records is not provided. However, a test for $x \in M$ is available

 x in M

with the result *false* or *true*.

 The simplest method to generate a set is the enumeration of the desired elements of the base type by

 [ElementList]

The empty set is member of any set type and is denoted by [ ].

Example 2.3.21:

 The values of the set type

  type digitset = **set of** 1..3;

 are the subsets

  [ ], [1], [2], [3], [1,2], [1,3], [2,3], and [1,2,3].

 Values of the character set

set of char

are for example

['a','b','c']   or   ['a'..'z', '0'..'9', 'ㄴ'].

The latter may also be denoted in form of a set expression:

['a'..'z'] + ['0'..'9'] + ['ㄴ'].

2.3.2.10   Files

A file consists of a sequence of arbitrarily many components of the same type. Therefore, the type definition of a file only fixes the type of the components:

| **file of** ComponentType

The component type may be any type except a file type or a dynamic array type.

The number of components in a file (the actual size of a file) is not determined by the definition of the file. It depends on the file operations applied to a file. Random access to the components of a file in the same manner as to the component variables of arrays and records is not available. Instead, a buffer variable of component type is provided which is declared automatically by the declaration of the corresponding file f. This buffer variable $f\uparrow$ facilitates the access to a special component of a file, called the actual component. The actual component is determined by the preceding file operations like *reset, rewrite, put,* or *get*:

rewrite(f)   initializes f for succeeding output operations. The first component of the file variable f is actual component. The procedure *rewrite* sets *eof(f)* = *true*, and the buffer variable $f\uparrow$ is undefined.

put(f)   assigns the value of $f\uparrow$ to the actual component, the following component becomes the actual component, and *eof(f)* = *true*. The buffer variable $f\uparrow$ is undefined.

reset(f)   initializes f for succeeding input operations. The first component becomes the actual component. If *eof(f)* = *true*, then the file is empty, and it is not possible to read anything. Thus, $f\uparrow$ is undefined. If *eof(f)* = *false*, i.e. the file is not empty, then the value of the actual component is assigned to the buffer variable.

get(f)   the succeeding component of the actual component becomes the new actual component. If *eof(f)* = *false*, then the value of the actual component is assigned to the buffer variable $f\uparrow$. Otherwise, the buffer variable $f\uparrow$ is undefined.

The logical function *eof* (*end of file*) returns *false* if the actual component is a defined component of the file. Otherwise the result is *true*.

Input and output using files are sequential processes beginning at the first component of the file. While reading, *eof(f)* = *false* must be valid. While writing, *eof(f)* = *true*.

Example 2.3.22:

```
program in_and_out;
var f : file of integer;
    ⋮
begin
  rewrite(f);              { Initialization for output }
  for i:= 1 to 100 do      { File f consists of }
  begin                    { 100 components with }
    f↑ := i;               { values 1 to 100 }
    put(f);
  end;
    ⋮
  reset(f);                { Initialization for input }
  while not eof(f) do      { The components of f are }
  begin
    writeln (sqr(f↑));     { read sequentially and }
    get(f);                { the squares of their values are printed }
  end;
end.
```

2.3.2.11   Text Files

A special file type is the predefined text file type *text* with the component type *char*. In principle, the handling of text files is the same as for other files. Since text files usually have a line structure, text files may additionally contain end-of-line characters which may be recognized by the logical function *eoln* (*end of line*).

If $eoln(t) = true$, the value of the actual component of the textfile variable t is the end-of-line character. In this case, the buffer variable $f\uparrow$ has the value ⊔ (blank).

The input/output procedures *read, readln, write,* and *writeln* simplify the handling of text file variables. They are used with a parameter for the file variable and with an input/output list (see section 2.5.2). If the file parameter is missing, the standard textfile variables *input* and *output* are assumed.

Example 2.3.23:

```
program make_a_copy (original, copy);
{ A text is copied according to its line structure }
var
  original, copy: text;
  ch: char;
    ...
begin
  reset (original);
  rewrite (copy);
  while not eof (original) do
```

```
    begin
        while not eoln (original) do
        begin
            read (original, ch);
            write (copy, ch);
        end;
        readln (original);
        writeln (copy);
    end;
end.
```

2.3.3 Structured Arithmetic Standard Types

-- *PASCAL–XSC* ----

PASCAL–XSC provides the additional arithmetic built-in types *complex, interval, cinterval, rvector, cvector, ivector, civector, rmatrix, cmatrix, imatrix,* and *cimatrix.* They have no constants and no operators within the language itself and therefore no expressions, either. The use of operators and standard functions requires the use of the corresponding arithmetic modules (see chapter 3).

2.3.3.1 The Type *complex*

-- *PASCAL–XSC* ----

The range of complex numbers

$$z = x + iy$$

with the real part x and the imaginary part y (i is the imaginary unit) may be declared in PASCAL by

type complex = **record** re, im: real **end**;

In PASCAL–XSC, this type is a predefined type. The variable declaration

var z: complex;

specifies a complex variable z. The real part and imaginary part of z may be accessed by $z.re$ and $z.im$, respectively.

2.3.3.2 The Type *interval*

--- PASCAL–XSC ---

For real intervals

$$a = [\underline{a}, \overline{a}] := \{x \in \mathbb{R} \mid \underline{a} \le x \le \overline{a}\}$$

with lower bound $\underline{a}$ and upper bound $\overline{a}$, PASCAL–XSC provides the predefined type *interval* declared by

type interval = **record** inf, sup: real **end**;

The variable declaration

var a: interval;

specifies an interval variable a. The lower and upper bound of a may be accessed by a.*inf* and a.*sup*, respectively:

a.inf   { access to lower bound }
a.sup   { access to upper bound }

2.3.3.3 The Type *cinterval*

--- PASCAL–XSC ---

Complex intervals are rectangles with sides parallel to the axes in the complex plane. The predefined type *cinterval* is defined by

type cinterval = **record** re, im : interval **end**;

The components of the complex interval variable c, declared by

var c: cinterval;

may be accessed by

c.re   { an interval for the real part }
c.im   { an interval for the imaginary part }

The component variables are intervals. The access to their real bounds may be done by

c.re.inf   c.re.sup   c.im.inf   c.im.sup   .

2.3.3.4 Vector Types and Matrix Types

```
                                                    PASCAL–XSC
```

For vectors and matrices with component type *real*, *complex*, *interval*, and *cinterval*, the following dynamic types are available:

```
type   rvector   =  dynamic array [*] of real;
       rmatrix   =  dynamic array [*] of rvector;
       cvector   =  dynamic array [*] of complex;
       cmatrix   =  dynamic array [*] of cvector;
       ivector   =  dynamic array [*] of interval;
       imatrix   =  dynamic array [*] of ivector;
       civector  =  dynamic array [*] of cinterval;
       cimatrix  =  dynamic array [*] of civector;
```

2.3.4 Pointers

All data types of ISO Standard PASCAL are static. Variables of these types are allocated at compile time. Their number remains unchanged during execution of the program. However, we frequently need to use a data structure which allows us to generate and discard variables as the need arises. For this purpose, the pointer type is provided.

A pointer variable p is a reference (the value is an address) to a variable $p \uparrow$ of the referenced type. This referenced variable $p\uparrow$ need not to be declared. It is generated by means of the standard procedure *new* during execution time of the program. The pointer variable itself is declared like any other static variable.

The type definition must specify only the referenced type:

type PointerTypeIdentifer = ↑ TypeIdentifier

The referenced type may be any Standard PASCAL type. In contrast to the principle that any quantity must be declared before it is used, the definition of the referenced type may follow the declaration of the pointer type. The values of a pointer type are references to variables of the referenced type extended by the value **nil** (pointer constant) referencing to no variable and belonging to any pointer type. The constant **nil** is the only value of a pointer type that is explicitly accessible.

Example 2.3.24:

```
type DateType     = array [1..20] of real;
     DatePointer  = ↑ element;
     element      = record date: DateType;
                           successor: DatePointer;
                    end;
```

New variables of the referenced type are allocated with the procedure *new*:

| new (PointerVariable);

Given the declarations in Example 2.3.24, then the statement

    new (DatePointer);

allocates a referenced variable *DatePointer*↑ of type *element*. The pointer variable *DatePointer* points to this referenced variable. The value of *DatePointer* is not explicitly known.

If the referenced type is a record type with variants, a particular variant may be allocated by

| new (PointerVariable, TagFieldValue);

Nested variants may be allocated by

| new (PointerVariable, TagFieldValue , · · ·, TagFieldValue);

The value of a pointer expression may be assigned to a pointer variable by

| PointerVariable := PointerExpression;

with the pointer expression being the constant **nil**, a pointer variable, or a function call with a result of pointer type. Functions with a result of pointer type are allowed. Since a pointer can point to any object, functions can return pointers to arbitrary types.

Pointer expressions may be compared by the relational operators $=$ and $<>$, e.g. $p = $ **nil** or $p <> q$.

When a dynamic pointer variable is no longer required by the program, the procedure

| dispose (PointerVariable);

is used to reclaim the memory occupied by the referenced variable. Afterwards, the value of the pointer variable and all references to the referenced variable are undefined. Referenced variables allocated by

    new (p, m1, m2, · · ·, mk);

must be released by

    dispose (p, m1, m2, · · ·, mk);

At the call of *dispose*, the values of *m1*, ... , *mk* must be identical with the corresponding values at the call of *new*.

Example 2.3.25:

```
var
    p : DatePointer;
begin
    ⋮
    new (p);
    p↑.date := { value corresponding to type DateType };
    p↑.successor := nil;
    work (p); { procedure call for further execution }
    ⋮
    dispose(p); { release of memory that is no longer required }
    ⋮
```

───────────────────────────────── *PASCAL–XSC* ─────

A referenced type may be any type except dynamic array types.

In addition to *dispose*, there is another method available to reclaim memory. The procedure call *mark (PointerVariable)* assigns the value of the heap pointer to the specified pointer variable. The subsequent procedure call *release (Pointer-Variable)* (with the same unchanged pointer variable as used with *mark*), sets the heap pointer to the address contained in its argument. The call *release (PointerVariable)* thus discards all dynamically allocated variables above this address. After this, the value of the pointer variable used is undefined, and all references to the released memory range are undefined.

Within a program and all used modules, a programmer may employ either the *dispose* construct or the *mark/release* construct, but not both.

2.3.5   Compatibility of Types

Certain operations are only executable if the types of the corresponding operands are compatible. Two types t_1 and t_2 are called *compatible* if

(a) t_1 and t_2 are the same type.

(b) t_1 is a subrange of t_2, t_2 is a subrange of t_1, or both t_1 and t_2 are subranges of the same base type.

(c) t_1 and t_2 both are set types of compatible base types, and both are either packed or unpacked.

(d) t_1 and t_2 are (static) string types with the same length.

Moreover, the *assignment compatibility* of the type t_1 of the variable on the left-hand side and the type t_2 of the expression on the right-hand side ($t_1 := t_2$) is defined by:

(a) t_1 and t_2 are the same type, except a file type.

(b) t_1 is type *real*, t_2 is type *integer*.

(c) t_1 and t_2 are compatible scalar types (except *real*), and the value of type t_2 is contained in t_1.

(d) t_1 and t_2 are compatible set types, and the elements of the value of type t_2 are contained in the base type of t_1.

(e) t_1 and t_2 are compatible (static) string types.

The assignment compatibility is also applied to the formal parameter in connection with a call by value of a function or procedure and the corresponding actual expression in case of a procedure or function call. A formal parameter for call by reference and the corresponding actual parameter must be compatible.

Example 2.3.26:

The declaration

```
type
    vec1 = array [1..10] of real;
    vec2 = array [1..10] of real;
    vec3 = vec1;
```

causes the types *vec1* and *vec3* to be compatible because they are the same type, whereas *vec1* and *vec2* are not compatible, although they have the same structure.

PASCAL–XSC

As a consequence of the dynamic types and the dynamic string concept, the compatibility must be extended, too. These extensions are explained in the following section.

Furthermore, a programmer may overload the assignment operator := (see section 2.7.12) in order to explicitly extend assignment compatibility to types which are otherwise not compatible. This "overloaded compatibility" is valid only for the assignment statement but not for the call by value of functions and procedures.

2.3.5.1 Compatibility of Array Types

PASCAL–XSC

As in Standard PASCAL, two array types are *compatible* only if they are the same types, i.e. a dynamic type is not compatible with a static type.

A value of the array type t_2 is *assignment compatible* with the variable of array type t_1 on the left-hand side, if

- both types are compatible, and the lengths of the corresponding index ranges are equal.

—————————————————————————— *PASCAL–XSC* ———

- t_1 is an *anonymous* type, and both are *structurally equivalent*.

  A variable is called of anonymous type if there is no related type identifier in the corresponding declaration. This may occur in the case of component variables (subarrays) (see section 2.3.2.2).

  Example 2.3.27:

  With the declarations

type	vector	=	**array** [1..5] **of** integer;
	v_matrix	=	**array** [1..5] **of** vector;
	a_matrix	=	**array** [1..5,1..5] **of** integer;
var	a	:	**array** [1..10] **of** real;
	b	:	**array** [1..10] **of** rvector[1..10];
	c	:	rmatrix[1..10, 1..10];
	d	:	vector;
	e	:	v_matrix;
	f	:	a_matrix;

  the variables (or component variables)

      a,   b,   c[*,2],   e[*,1],   and   f[3]

  are of anonymous type, and the variables (or component variables)

      b[3],   c[2],   d,   e[1],   and   f

  are of known type.

Two array types are called structurally equivalent if the component types are the same and the index ranges are identical in number, length, and base type. If the index ranges of an array type are not yet specified, the length is always adequate. This is a special case for a formal parameter.

Thus, the assignment statement (see section 2.5.1) is allowed in the following cases:

Type of left Side	Type of right Side	Assignment permitted
anonymous dynamic	arbitrary array type	if structurally equivalent
known dynamic	known dynamic	if the same type
anonymous static	arbitrary array types	if structurally equivalent
known static	known static	if the same type

In all other cases, an assignment is only possible by *qualification* of the array expression of the right side:

| ArrayTypeIdentifier (ArrayExpression)

─────────────────────────── *PASCAL–XSC* ───────

In this case, the array type identifier serves as type conversion function (see section 2.4.3.1). Qualification, however, requires the named type and the type of the array expression to be structurally equivalent.

Example 2.3.28:

The types *poly* and *vec* declared by

> **const** degree = . . . ;
> **type** poly   = **dynamic array** [∗] **of** real;
>   vec   = **dynamic array** [∗] **of** real;

are not compatible. If we have provided a vector addition operator for the type *vec*, then the polynomials p and q declared by

> **var** p, q: poly[0..degree];

can be added by means of the qualification

> p := poly (vec (p) + vec (q)).

2.3.5.2 Compatibility of Strings

In Standard PASCAL, string types (called array-string types in the following) are compatible and assignment compatible only if their lengths coincide.

─────────────────────────── *PASCAL–XSC* ───────

The following rules apply to the new standard type *string* (called *string* type in the following):

- Two *string* types are always *compatible*. A *string* type, however, is not *compatible* with any other type.

- A string value of type t_2 is *assignment compatible* with a variable of type t_1 if t_1 is a *string* type and t_2 is an *array-string* type, a *string* type, or a *char* type.

2.4   Expressions

In this section, we describe the expression concept of PASCAL. For the additional types of PASCAL–XSC, we supply the corresponding details. Moreover, we describe how to create expressions for arbitrary, user-defined types by declaring operators and functions with arbitrary result for these types. This user-defined expression concept is processed according to the usual rules of priority and parenthesizing.

2.4.1   Standard Expressions

Expressions for the types *integer*, *real*, *boolean*, *char*, enumeration type, and *set* are composed of operands and operators in the usual manner. All rules and properties described for the types *integer*, *boolean*, *char*, and enumeration type in the following apply in the same way to their subrange types.

The evaluation of an expression is done according to the conventional rules of algebra for left-to-right evaluation of operators and operator precedence. An expression enclosed within parentheses is evaluated independently of preceding and following operators. The type of the expression value is given by the operator which is processed last. An expression is built up by

> MonadicOperator
> > Operand   { not empty }
> > > DyadicOperator   Operand   ...

An operand is given by the alternatives

- constant
- variable
- function call
- expression, enclosed in parentheses

where a function can be predefined or user-defined.

Example 2.4.1:

> Let *op1*, *op2*, *op3* be operands, − a monadic operator, and +, * dyadic operators. Then, we can built up the expression
>
> > − op1 + op2 * op1 * op3
>
> using three repetitions of the third line of the above syntax. Moreover, we can replace *op3* by a further expression, for example
>
> > op2 + f (op1)
>
> where f is a function with appropriate result type. Then we get
>
> > − op1 + op2 * op1 * (op2 + f (op1))

The operators are defined only for special kind of operands. They denote different operations depending on the operand types:

Monadic Operator	Operand Type	Result Type
+, −	*integer, real*	*integer, real*
not	*boolean*	*boolean*

Dyadic Operator	Operand Type	Result Type
+, −, *, **div**, **mod** /	*integer*	*integer* *real*
+, −, *, /	*integer* and *real*	*real*
+, −, *, /	*real*	*real*
or, and	*boolean*	*boolean*
+ (set union), − (set difference), * (set intersection)	*set*	*set*
=, <>, <, >, <=, >=	*integer, real, char, boolean, enumeration type, string*	*boolean*
=, <>, <= (subset inclusion), >= (superset inclusion)	*set*	*boolean*
in (set membership)	left operand: *integer, boolean*, and enumeration type right operand: corresponding set type	*boolean*

The priority levels of Standard PASCAL are:

Priority	Dyadic Operators	Monadic Operators
0 (lowest)	=, <>, <=, >=, <, >, **in**	
1	+, −, **or**	+, −
2	**mod, div**, *, /, **and**	
3 (highest)		**not**

PASCAL–XSC

In contrast to Standard PASCAL, the monadic operators + and − have the highest priority 3. Several monadic operators can occur in sequence.

MonadicOperator ...

Operand { not empty }

DyadicOperator   Operand   ...

--- PASCAL–XSC ---

Several monadic operators in sequence are executed from right to left. In addition to Standard PASCAL, the following operators are available:

Dyadic Operator	Operand Type	Result Type
+<, −<, *<, /<	*integer* or *real*	*real*
+>, −>, *>, />	*integer* or *real*	*real*
+	*char* or *string*	*string*
in	left operand: *string* or *char* right operand: *string*	*boolean*

Additionally, the dyadic operator symbols **, +*, and >< are available. They get their predefined meaning by using the arithmetic modules (see chapter 3). In PASCAL–XSC, the priority levels are:

Priority	Dyadic Operators	Monadic Operators
0 (lowest)	=, <>, <=, >=, <, >, in, ><	
1	or, +, +<, +>, −, −<, −>, +*	
2	*, *<, *>, /, /<, />, ** mod, div, and	
3 (highest)		+, −, not

2.4.1.1 Integer Expressions

An *integer* expression is composed of *integer* operands and the operators $+, -, *$, **div**, and **mod**. The operators **div** and **mod** denote the *integer* division and the division remainder, respectively. The following intrinsic functions are available:

Function	Definition
trunc (*real* expression)	Rounding by truncation of the fractional part
round (*real* expression)	Rounding to the nearest *integer* number, i.e. $round\ (r) = \begin{cases} trunc\ (r + 0.5) & \text{for } r \geq 0 \\ trunc\ (r - 0.5) & \text{for } r < 0 \end{cases}$
ord (O-Type expression)	Ordinal number of the parameter. The elements of these types have the corresponding ordinal numbers 0,1,2,...
ord (*integer* expression)	Identity, i.e. $ord(v) = v$
succ (*integer* expression)	Successor, i.e. $succ(v) = v + 1$
pred (*integer* expression)	Predecessor, i.e. $pred(v) = v - 1$
abs (*integer* expression)	Absolute value
sqr (*integer* expression)	Square, i.e. $sqr(v) = v^2$

O-Type = *boolean, char,* or *enumeration type*

—————————————————————— PASCAL–XSC ——————

The following additional functions with *integer* result are available:

Function	Definition
loc (variable)	Implementation-dependent address of the variable
ord (pointer expression)	Implementation-dependent value of the pointer expression
sign (S-Type expression)	Sign, i.e. $sign\ (a) = \begin{cases} -1 & \text{for } a < 0 \\ 0 & \text{for } a = 0 \\ +1 & \text{for } a > 0 \end{cases}$
lbound (array variable, *integer* constant)	Lower bound of an index range
lb (array variable, *integer* constant)	Lower bound of an index range
ubound (array variable, *integer* constant)	Upper bound of an index range
ub (array variable, *integer* constant)	Upper bound of an index range
expo (*real* expression)	Exponential part of the normalized mantissa (see section 2.4.1.2)

S-Type = *integer, real,* or *dotprecision*

For an array variable A, the function *lbound(A,n)* delivers the lower bound of the index range of the n-th dimension. If there is no second parameter, the first dimension is chosen. The functions *lb* (for *lbound*) and *ub* (for *ubound*) can be used as short forms.

The function

    ival

for conversion of a string to an *integer* value (see section 2.9) is provided.

    An *integer* expression may also include user-defined operators and function calls with *integer* result type.

2.4.1.2   Real Expressions

A *real* expression is composed of *real* or *integer* operands and the corresponding floating point operators $+, -, *, /$. Using $+, -, *$ with two *integer* operands causes the *integer* operation to be executed. The following predefined functions are available:

Function	Definition
abs (*real* expression)	Absolute value $\quad |x|$
sqr (*real* expression)	Square $\quad x^2$
sin (*real* expression)	Sine $\quad \sin x$
cos (*real* expression)	Cosine $\quad \cos x$
arctan (*real* expression)	Arc Tangent $\quad \arctan x$
exp (*real* expression)	Exponential Function e^x
ln (*real* expression)	Natural Logarithm $\ln x$, $x > 0$
sqrt (*real* expression)	Square Root $\sqrt{x}$, $x \geq 0$

Further implementation-dependent information about the domain and the range of the functions can be found in the user manual of the compiler.

Example 2.4.2:

With the declarations

```
var   x, y, v, w   :   real;
      i, j         :   integer;
```

the expressions

sqr(x) + sin(y+1.5)/ln(sqr(v)+sqr(w)+1.2)   and
i **div** j + 1e−10

are *real* expressions.

─────────────────────────────── *PASCAL–XSC* ───

PASCAL–XSC provides floating point operations with three different kinds of roundings. The following remarks give a review of the fundamentals for the use of these operations.

A floating point system R is characterized by a base b (for instance 2 or 10), a finite number n of mantissa digits (for instance 13), and an exponent range with the smallest exponent *emin* and the largest exponent *emax* (see also chapter 1). A normalized floating point number x can be represented by

$$x = \pm 0.d_1 d_2 ... d_n \cdot b^{ex},$$

where $d_1 \neq 0$, $0 \leq d_i \leq b - 1$, and $emin \leq ex \leq emax$. We denote a floating point system by $R = R(b, n, emin, emax)$.

A floating point system (see also [28] and [24]) is not closed with respect to the arithmetic operations $+, -, *, /$. That means that the mathematical operation applied to two operands in R does not always produce a result which lies in R. Using $R(10, 2, -10, 10)$ for example, $x + y$ with $x = 0.58$ and $y = 0.47$ delivers 1.05. This number is not an element of R, so it must be rounded to a number in R. The best we can do is to round the exact result to one of the adjacent numbers in R, either 1.0 or 1.1. So the result of the rounded operation is correct up to one ulp (<u>u</u>nit in the <u>l</u>ast <u>p</u>lace).

PASCAL–XSC

The smallest local error is produced by using the rounding to the nearest floating point number (1/2 ulp). In PASCAL–XSC, this implementation-dependent rounding is accessed by the usual operations $+, -, *, /$. The operations with downwardly directed rounding and with upwardly directed rounding are denoted by the symbols $+<, -<, *<, /<$, and $+>, ->, *>, />$, respectively.

We need directed roundings if we want to compute guaranteed bounds for the exact value of a *real* expression. To get a valid bound, we must be careful to use the correct rounding mode for each operation. We must also take care to round literal constants correctly.

Directed roundings are also used to implement an interval arithmetic. In each interval operation, the lower bound must be rounded downward, and the upper bound must be rounded upward.

Example 2.4.3:

In $R(10, 4, -5, 5)$, the *real* expressions

$$1/3, \; 1/<3, \text{ and } 1/>3$$

deliver the values

$$0.3333, \; 0.3333, \text{ and } 0.3334,$$

respectively.

If we want to compute a lower and an upper bound for the real expression

$$x \cdot y - v \cdot w,$$

we can do this in PASCAL–XSC by evaluating

$$x *< y -< v *> w$$

and

$$x *> y -> v *< w \, .$$

The result of $v \cdot w$ must be rounded the opposite direction as the result of $x \cdot y$ because of the intervening subtraction operator.

PASCAL–XSC provides the built-in functions *succ* and *pred* for both *integer* and *real* arguments.

Function	Definition
succ (*real* expression)	Next larger floating point number
pred (*real* expression)	Next smaller floating point number

PASCAL–XSC

The following additional mathematical functions are provided:

Function	Definition
exp2 (*real* expression)	Power function, base 2 2^x
exp10 (*real* expression)	Power function, base 10 10^x
log2 (*real* expression)	Logarithm, base 2 $\log_2 x$
log10 (*real* expression)	Logarithm, base 10 $\log_{10} x$
tan (*real* expression)	Tangent $\tan x$
cot (*real* expression)	Cotangent $\cot x$
arcsin (*real* expression)	Arc Sine $\arcsin x$
arccos (*real* expression)	Arc Cosine $\arccos x$
arccot (*real* expression)	Arc Cotangent $\text{arccot}\, x$
arctan2 (*real* expression, *real* expression)	$\text{arctan2}(r1,r2) = \arctan(r1/r2)$
sinh (*real* expression)	Hyperbolic Sine $\sinh x$
cosh (*real* expression)	Hyperbolic Cosine $\cosh x$
tanh (*real* expression)	Hyperbolic Tangent $\tanh x$
coth (*real* expression)	Hyperbolic Cotangent $\coth x$
arsinh (*real* expression)	Inverse Hyperbolic Sine $\text{arsinh}\, x$
arcosh (*real* expression)	Inverse Hyperbolic Cosine $\text{arcosh}\, x$
artanh (*real* expression)	Inverse Hyperbolic Tangent $\text{artanh}\, x$
arcoth (*real* expression)	Inverse Hyperbolic Cotangent $\text{arcoth}\, x$

All *real* arithmetic functions available in PASCAL–XSC deliver a result of maximum accuracy in the sense that there is no other floating-point number between the exact result and the computed floating-point number (1 ulp accuracy). Further implementation-dependent information about the domain and the range of the functions can be found in the user manual of the compiler.
The function

  rval

converts strings into *real* values (see section 2.9 for details). For decomposing and composing of *real* numbers, PASCAL-XSC provides the functions *mant* and *comp* (see also *expo* in section 2.4.1.1).

Function	Definition
mant (*real* expression)	Normalized mantissa *m* of *r*. The range of *m* is implementation-defined.
comp (*real* expression, *integer* expression)	Composition of a mantissa (type R) and an exponent (type I) to a *real* number. The ranges of the *real* and *integer* expressions are implementation-defined.

```
                                                    PASCAL–XSC
```

Example 2.4.4:

The functions *mant*, *expo*, and *comp* satisfy the identities

$$x = comp\ (\ mant\ (x),\ expo\ (x)\ ),$$
$$e = expo\ (\ comp\ (m\ ,\ e)\ ),$$
$$m = mant\ (\ comp\ (m\ ,\ e)\ ).$$

Depending on the implementation, we might get

Statement			Result		
m	:=	mant (100)	m	=	0.1
e	:=	expo (100)	e	=	3
x	:=	comp (m,e)	x	=	100 = 0.1E+03

A *real* expression may also include user-defined operators and function calls with *real* result type.

2.4.1.3   Boolean Expressions

Permissible operands in a *boolean* expression are the literal constants *true* and *false*, variables, *boolean* functions, comparisons, expressions in parentheses, and the following *boolean* functions:

Function	Definition
pred (*boolean* expression)	Predecessor according to *false* < *true*
succ (*boolean* expression)	Successor according to *false* < *true*
odd (*integer* expression)	Returns *true* if the argument is an odd number, and *false* if it is an even one.
eof (file variable)	Returns *true* if the end of the file is reached, and *false* if not.
eoln (text file variable)	Returns *true* if the end of the line is reached, and *false* if not.

If one of the operands is a comparison, then it has to be put in parentheses. The symbol <> stands for $\neq$ (not equal to). The relational operators <= and >= denote the logical implication $\rightarrow$ and $\leftarrow$, respectively. The symbol = denotes the logical equivalence.

─────────────── *PASCAL–XSC* ───

Additional built-in functions:

Function	Definition
lbound (array variable, *integer* constant)	Lower bound of an index range
lb (array variable, *integer* constant)	Lower bound of an index range
ubound (array variable, *integer* constant)	Upper bound of an index range
ub (array variable, *integer* constant)	Upper bound of an index range

For an array variable A with index type *boolean*, *lbound(A,n)* delivers the lower bound of the index range of the n-th dimension. If there is no second parameter, the first dimension is chosen. This rule also holds for *ubound*. *lb* (for *lbound*) and *ub* (for *ubound*) can be used as short forms.

Comparisons for values of the arithmetic types

> *complex, interval, cinterval*

and

> *rvector,   cvector,   ivector,   civector,*
> *rmatrix,   cmatrix,   imatrix,   cimatrix*

are defined in the corresponding arithmetic modules. A detailed description is given in chapter 3 (Arithmetic Modules).
   It is not possible to compare *dotprecision* values directly. Two *dotprecision* values can be compared by subtracting them and then using the *sign* function:

$$\text{sign} (d) := \begin{cases} 1 & \text{for} \quad d > 0 \\ 0 & \text{for} \quad d = 0 \\ -1 & \text{for} \quad d < 0, \end{cases}$$

where d is an expression of type *dotprecision*.
   A *boolean* expression may also include user-defined operators and function calls with *boolean* result type.

2.4.1.4   Character Expressions

A character expression is given by a constant, a variable, or a function call. There are no character operators. Predefined functions with result type *char* are:

Function	Definition
pred (*char* expression)	Predecessor
succ (*char* expression)	Successor
chr (*integer* expression)	Returns the character with the ordinal value of the *integer* expression

The results of these *char* functions depend on the implementation.

PASCAL–XSC

Additional predefined functions:

Function	Definition
lbound (array variable, *integer* constant)	Lower bound of an index range
lb (array variable, *integer* constant)	Lower bound of an index range
ubound (array variable, *integer* constant)	Upper bound of an index range
ub (array variable, *integer* constant)	Upper bound of an index range

For an array variable A with its index type *char*, *lbound(A,n)* delivers the lower bound of the index range of the n-th dimension. If there is no second parameter, the first dimension is chosen. This rule also holds for *ubound*. *lb* (for *lbound*) and *ub* (for *ubound*) can be used as short forms.

A *char* expression may also include user-defined operators and function calls with *char* result type.

2.4.1.5 Enumeration Expressions

The enumeration expression consists of enumeration constants, variables, and function calls of the built in functions *pred* and *succ*.

Function	Definition
pred (enumeration expression)	Predecessor in the enumeration type
succ (enumeration expression)	Successor in the enumeration type

---

PASCAL–XSC

Additional predefined functions:

Function	Definition
lbound (array variable, *integer* constant)	Lower bound of an index range
lb (array variable, *integer* constant)	Lower bound of an index range
ubound (array variable, *integer* constant)	Upper bound of an index range
ub (array variable, *integer* constant)	Upper bound of an index range

For an array variable A with an enumeration type as its index type, *lbound(A,n)* delivers the lower bound of the n-th dimension. If there is no second parameter, the first dimension is chosen. This rule also holds for *ubound*. *lb* (for *lbound*) and *ub* (for *ubound*) can be used as short forms.

──────────────────────────────── *PASCAL–XSC* ────

Example 2.4.5:

    type precipitation = (rain, hail, snow);
    var p: precipitation;
        pset: **array** [precipitation] **of** real;

        ⋮

    p:= succ (rain); { the value hail is assigned to p }
    p:= ubound (pset); { the value snow is assigned to p }

An *enumeration* expression may also include user-defined operators and function calls with *enumeration* result type. There are no predefined operators available.

2.4.2   Accurate Expressions (#-Expressions)

──────────────────────────────── *PASCAL–XSC* ────

The usual *real* expressions of almost every programming language are simply evaluated by executing each operation and immediately rounding the result to the given *real* format. The problem with this kind of evaluation is that the influence of the roundings may falsify the final result. To avoid such uncontrollable effects, PASCAL–XSC provides the *dotprecision* expressions and accurate expressions. Accurate expressions (#-expressions) are marked by the preceding #-symbol. There are three different forms of basic accurate expressions: the *dotprecision* expression

> **▌**   # (*real* ExactExpression) { exact *dotprecision* result }

the *real* accurate expression

> **▐**   #∗ (*real* ExactExpression) { rounded to the nearest real number   }
>   #< (*real* ExactExpression) { rounded to the next smaller real number }
>   #> (*real* ExactExpression) { rounded to the next larger real number }

and the *interval* accurate expression

> **▌**   ## (*real* ExactExpression) { rounded to the smallest enclosing *interval* }

The exact expression enclosed in parentheses is always evaluated exactly without any rounding. An exact expression must be mathematically equivalent to a scalar product (dot product) $\sum u_i \cdot v_i$. It is built up according to the following syntactic structure:

——————————————— *PASCAL–XSC* ———

± Summand

        Operator   Summand ...

Only the operators + and − can be used in the exact expression. They denote
the exact (errorless) addition and subtraction of operands (summands) of one or
more of the following forms:

Summand	Definition
dotprecision variable	*dotprecision*-variable
real operand	*real* operand with the alternatives
	integer variable *integer* constant *real* variable *real* constant
real operand ∗ *real* operand	exact double length product of two *real* operands
(*real* exact expression)	exact expression enclosed in parentheses
for i:= s **to** e **sum** (*real* exact expression)	**for**-statement for summation, with *i* an *integer* variable and *s, e integer* expressions
for i:= s **downto** e **sum** (*real* exact expression)	**for**-statement for summation, with *i* an *integer* variable and *s, e integer* expressions

Notice: Within the exact expression, the operators +, −, ∗ denote the exact
operations in the mathematical sense without any rounding. Therefore,
they can <u>not</u> be overloaded by user-defined operators.

The **for**-statement with **sum** is a short form for summation. In this statement, the exact expression enclosed in parentheses may depend on the
control variable *i*. An expression of the form

$$EE_s + EE_{s+1} + ... + EE_{e-1} + EE_e$$

with exact expressions EE_i can be abbreviated by

    for $i:= s$ **to** e **sum** (EE_i)

(see also section 2.5.8.3 **for**-statement).

-- *PASCAL–XSC* --------

An empty loop (**for**-statement) corresponds to a summand with the value zero. The same applies to **downto**.

The *integer* expressions s (start index) and e (end index) themselves must not contain explicit #-expressions.

Example 2.4.6:

The value of the scalar product

$$s := \sum_{i=1}^{10} a_i * b_i$$

can be computed with only one rounding. Assuming the declarations

```
var   a, b  :   array [1..10] of real;
      s     :   real;
      d     :   dotprecision;
      i     :   integer;
```

this can be realized via

```
d := # (0);
for i:=1 to 10 do  d := # (d + a[i]*b[i]);
s := #* (d); { rounding to the nearest real number }
```

Using the short form, this can be done by

```
s := #* (for i:=1 to 10 sum (a[i]*b[i]));
```

Example 2.4.7:

To compute the nearest, the next smaller, and the next larger floating point number of the value of the expression $E = x * y - v * w$, we can write

$$E_{nearest} := \#* (x * y - v * w),$$
$$E_{smaller} := \#< (x * y - v * w), \text{ and}$$
$$E_{larger} := \#> (x * y - v * w),$$

respectively. The results satisfy

$$\text{pred} (E_{larger}) = E_{smaller} \le E_{nearest} \le E_{larger} = \text{succ} (E_{smaller}).$$

Notice: If literal constants are used as *real* operands within an #-expression, the programmer should understand that these constants are converted into the internal data format first. Thus, depending on the implementation, inevitable errors may arise with the necessary conversions.

PASCAL–XSC

For example, with an internal binary representation, the expression

   ## (0.1)

does not deliver an interval inclusion of the real number 0.1, but a point interval corresponding to the value of the converted constant. An inclusion for the real value 0.1 may be computed by

   intval ((<0.1) , (>0.1))

(see also section 3.2).

2.4.3   Expressions for Structured Types and Pointer Expressions

The set type is the only structured type of Standard PASCAL for which expressions can be built up in the usual manner with operations. There are no operators in array or record expressions. There are no file expressions or text file expressions.

PASCAL–XSC

The operator concept of PASCAL–XSC (see section 2.7.6) enables us to declare operators for arbitrary predefined types and user-defined types. Thus, we can define expressions of any type.

The syntax of a general expression in PASCAL–XSC is identical to the syntax described in section 2.4.1 on page 45 for standard expressions.

PASCAL–XSC provides expressions for the arithmetic types *complex, interval, cinterval, rvector, cvector, ivector, civector, rmatrix, cmatrix, imatrix,* and *cimatrix.* It also provides many operators and functions for these types. A detailed description of these features is given in chapter 3 (Arithmetic Modules) in the corresponding sections describing the modules C_ARI, I_ARI, CI_ARI, MV_ARI, MVC_ARI, MVI_ARI, and MVCI_ARI.

2.4.3.1   Array Expressions

An array expression comprises no operators. It consists only of variables.

PASCAL–XSC

In PASCAL–XSC, an array expression can be composed of user-defined operators, variables, function calls and *qualification* (similar to the casting in C).

PASCAL–XSC

The qualification has the form

❚   ArrayTypeIdentifier (ArrayExpression)

where the array type identifier serves as a type converting function. If the array expression is structurally equivalent, then it is converted into the type named by the identifier.

There are no predefined operators for operands of an array type.

Example 2.4.8:

```
type
  vector = array [1..8] of real;
  polynomial = array [0..7] of real;
var
  v : vector;
  p : polynomial;
...
  p := polynomial (v);
  v := vector (p);
```

The type converting function or qualification is used in connection with dynamic arrays and with operators (see section 2.7.6).

A dynamic array expression has the same syntactical structure as the array expression, except that dynamic array operands can be used.

2.4.3.2 String Expressions

There are no predefined operators or functions for strings in Standard PASCAL. A string expression is either a string constant or a string variable.

PASCAL–XSC

The operator + defined by

operator + (a, b : string) conc: string;

concatenates two dynamic string operands. The strings are concatenated in the order a followed by b. The current length of the result is the sum of the current lengths of a and b. If the maximum length of the type *string* is exceeded, then the result is implementation-defined.

PASCAL-XSC ──

The operands of + may be *string* constants and variables, *string* function calls (for the predefined functions, see section 2.9), and *string* expressions enclosed in parentheses. A character expression can be used as special kind of a *string* operand.

Example 2.4.9:

        var s1, s2: string [6];
            s3 : string [11];

            ⋮

        s1 := 'PASCAL';
        s2 := '–XSC';
        s3 := s1 + s2; { the value 'PASCAL–XSC' is assigned to s3 }

2.4.3.3 Record Expressions

PASCAL-XSC ──

A record expression may include user-defined operators, record variables, and function calls with record result type. There are no predefined operators available.

2.4.3.4 Set Expressions

A set can be given by a set constructor of the form

❙   [ExpressionList]

The expressions in the expression list are element specifications. Such an element specification is an expression of the base type of the set or a subrange expressed by

        Expression .. Expression.

An empty expression list is permitted, so that [ ] defines an empty set.
    Set constructors, set variables, set function calls, and set expressions enclosed in parentheses can be operands in set expressions.
    The operators +, −, and ∗ denote the set union, set difference, and set intersection, respectively.

Example 2.4.10:

        var set_of_vowels, set_of_consonants : **set of** 'a'..'z';

        set_of_vowels := ['a', 'e', 'i', 'o', 'u'];
        set_of_consonants := ['a'..'z'] − set_of_vowels;

2.4.3.5 Pointer Expressions

A pointer expression consists of the constant nil, a pointer variable, or a pointer function call. There are no predefined operators for pointer operands.

PASCAL–XSC

If the function *ord* is applied to an argument of pointer type, it delivers the value of the pointer, i.e. the implementation-dependent address of the object which the pointer references. If p is a pointer, then

> ord (p) = loc (p↑)

2.4.4   Extended Accurate Expressions (#-Expressions)

PASCAL–XSC

The concept of *real* accurate expressions (#-expressions) based upon the type *dotprecision* can be extended to the arithmetic types *complex* , *interval*, and *cinterval* using the predefined operators +, −, and ∗. Moreover, it is possible to form accurate expressions for vectors and matrices over the types *real*, *interval*, *complex*, and *cinterval*. The corresponding exact expressions must be mathematically equivalent to scalar products (dot products).

Notice: To use these extended #-expressions, it is necessary to include the corresponding arithmetic module (see section 3) via a **use**-clause.

Within the exact expression, the operators +, −, and ∗ denote the exact operations in the mathematical sense without any rounding. Therefore, they can <u>not</u> be overloaded by user-defined operators.

The **for**-statement with **sum** can be used within the extended #-expression. An expression of the form

$$EE_s + EE_{s+1} + ... + EE_{e-1} + EE_e$$

where the EE_i are exact expressions, can be abbreviated by

> **for** $i:= s$ **to** e **sum** (EE_i)

(see also section 2.5.8.3, **for** statement). An empty loop (**for**-statement) corresponds to a summand with the value zero. The same applies to **downto**. The *integer* expressions s and e themselves must not contain #-expressions.

2.4.4.1 #-Expressions for the Arithmetic Types

─────────────────────────────────────── *PASCAL–XSC* ───────

A *real* accurate expression can use scalar products of two *real* vectors as summands. For example, with *a* and *b* of type *rvector*, we can evaluate *a* ∗ *b* and store the exact result in a *dotprecision* variable. Within the accurate expression, the operator ∗ which is provided in the arithmetic module MV_ARI denotes the exact computation of the scalar product.

Furthermore, accurate expressions are useful to program operations on the types *interval*, *complex*, and *cinterval* (*complex interval*). The syntax of accurate expressions of type *interval*, *complex*, and *cinterval* has the following form:

The *interval* accurate expression (with rounding to the smallest enclosing interval):

> ## (*real* ExactExpression)
> ## (*interval* ExactExpression)

The *complex* accurate expression (with componentwise rounding to the nearest, the next smaller, or the next larger *complex* number):

> #∗ (*complex* ExactExpression)
> #< (*complex* ExactExpression)
> #> (*complex* ExactExpression)

The *cinterval* accurate expression (with rounding to the smallest enclosing interval):

> ## (*complex* ExactExpression)
> ## (*cinterval* ExactExpression)

The exact expressions within these accurate expressions are of the same syntactical structure as the *real* exact expression, except that there are no *interval*-, *complex*-, or *cinterval-dotprecision* types or variables. In general, an exact expression has the syntactical form

> ± Summand
> 　　　　Operator Summand ...

where only + and − are permitted as operators.

────────────────────────────────────── *PASCAL–XSC* ──────

Summands which can be combined by the operators + and − are (with $\tau, \sigma \in$ {*real, interval, complex, cinterval*}):

Summand	Definition
dotprecision variable	*dotprecision* variable
τ operand	constant, variable, function call
τ operand ∗ σ operand	exact product of double length
τ *vector* operand ∗ *vector* σ operand	exact scalar product of two vectors
(τ exact expression)	exact expression enclosed in parentheses
for i:= s **to** e **sum** (τ exact expression)	**for**-statement for summation, with *i* an *integer* variable and *s,e integer* expressions
for i:= s **downto** e **sum** (τ exact expression)	**for**-statement for summation, with *i* an *integer* variable and *s,e integer* expressions

Not all the summands must be of the same type. Within a *cinterval* accurate expression, mixed summands of type *real, complex,* or *interval* can be used as well. The type of the exact expression is specified by the type combination of the summands that occur. The allowed τ- or σ-operands are listed in section 2.4.4.4.

Example 2.4.11:

Assuming the declarations

```
var  a, b    : real;
     ca      : complex;
     cib     : cinterval;
     v, w    : rvector[1..10];
     cv, cw  : cvector[1..10];
     civ     : civector[1..10];
```

the following accurate expressions are syntactically correct:

Acccurate Expression	Result Type
#< (b + v ∗ w + **for** i:=1 **to** 10 **sum** (v[i]))	real
#∗ (ca + a ∗ b + a ∗ ca + cv ∗ w + cv[3] ∗ cw[5])	complex
## (b + a ∗ b + v ∗ w)	interval
## (ca + a ∗ b + ca ∗ cib + cv ∗ civ)	cinterval

2.4.4.2 #-Expressions for Vectors

PASCAL–XSC ——

For vectors over the arithmetic types *real*, *complex*, *interval*, and *cinterval*, accurate expressions can be formed analogously:

Accurate Expression Type	Syntax	
rvector accurate expression	#*	(*rvector* ExactExpression)
	#<	(*rvector* ExactExpression)
	#>	(*rvector* ExactExpression)
ivector accurate expression	##	(*rvector* ExactExpression)
	##	(*ivector* ExactExpression)
cvector accurate expression	#*	(*cvector* ExactExpression)
	#<	(*cvector* ExactExpression)
	#>	(*cvector* ExactExpression)
civector accurate expression	##	(*cvector* ExactExpression)
	##	(*civector* ExactExpression)

The exact expression has the form

± Summand
     Operator   Summand   ...

where only + and − are permitted as operators.

Summands which can be combined by the operators + and − are (with $\tau, \sigma \in$ {*real*, *interval*, *complex*, *cinterval*}):

Summand	Definition
τ *vector* operand	variable, function call
τ operand * σ *vector* operand	exact product of double length (componentwise)
τ *vector* operand * σ operand	exact product of double length (componentwise)
τ *matrix* operand * σ *vector* operand	exact matrix/vector product (with exact scalar product for each component)
(τ *vector* exact expression)	exact expression enclosed in parentheses
for i:= s **to** e **sum** (τ *vector* exact expression)	**for**-statement for summation, with *i* an *integer* variable and *s, e integer* expressions
for i:= s **downto** e **sum** (τ *vector* exact expression)	**for**-statement for summation, with *i* an *integer* variable and *s, e integer* expressions

─────────────────────────────── *PASCAL–XSC* ───

Not all the summands must be of the same type. Within a *civector* accurate expression, mixed summands of type *rvector*, *cvector*, or *ivector* can be used as well. The type of the exact expression is specified by the type combinations of the summands that occur. The allowed τ- or σ-operands are listed in section 2.4.4.4.

Example 2.4.12:

Assuming the declarations

var	a, b	: real;		M	: rmatrix[1..10,1..10];
	ca	: complex;		cM	: cmatrix[1..10,1..10];
	v, w	: rvector[1..10];		iM	: imatrix[1..10,1..10];
	cv	: cvector[1..10];		ciM	: cimatrix[1..10,1..10];
	civ	: civector[1..10];			

the following accurate expressions are syntactically correct:

Accurate Expression	Result Type
#* (**for** i:=1 **to** 10 **sum** (M * v + a * M[*,i]))	rvector
#> (cv + v * b + a * cv + cM * w + cM * cv)	cvector
## (v + a * v + iM * v)	ivector
## (cv + M * v + ca * civ + ciM * cv)	civector

2.4.4.3   #-Expressions for Matrices

─────────────────────────────── *PASCAL–XSC* ───

For the arithmetic matrix types, accurate expressions can be formed analogously:

Accurate Expression Type	Syntax	
rmatrix accurate expression	#*	(*rmatrix* ExactExpression)
	#<	(*rmatrix* ExactExpression)
	#>	(*rmatrix* ExactExpression)
imatrix accurate expression	##	(*rmatrix* ExactExpression)
	##	(*imarrix* ExactExpression)
cmatrix accurate expression	#*	(*cmatrix* ExactExpression)
	#<	(*cmatrix* ExactExpression)
	#>	(*cmatrix* ExactExpression)
cimatrix accurate expression	##	(*cmatrix* ExactExpression)
	##	(*cimatrix* ExactExpression)

```
────────────────────────────── PASCAL-XSC ──────────
```

The exact expression again has the form

    ± Summand
                     Operator   Summand   ...

where only + and − are permitted as operators.

Summands which can be combined by the operators + and − are (with $\tau, \sigma \in$ {*real, interval, complex, cinterval*}):

Summand	Definition
τ *matrix* operand	variable, function call
τ operand $*$ σ *matrix* operand	exact product of double length (componentwise)
τ *matrix* operand $*$ σ operand	exact product of double length (componentwise)
τ *matrix* operand $*$ σ *matrix* operand	exact matrix product (with exact scalar product for each component)
(τ *matrix* exact expression)	exact expression enclosed in parentheses
for i:= s **to** e **sum** (τ *matrix* exact expression)	**for**-statement for summation, with *i* an *integer* variable and *s, e integer* expressions
for i:= s **downto** e **sum** (τ *matrix* exact expression)	**for**-statement for summation, with *i* an *integer* variable and *s, e integer* expressions

Not all the summands have to be of the same type. Within a *cimatrix* accurate expression, mixed summands of type *rmatrix*, *cmatrix*, or *imatrix* can be used as well. The type of the exact expression is specified by the type combination of the summands that occur. The allowed τ-or σ-operands are listed in section 2.4.4.4.

Example 2.4.13:

Assuming the declarations of example 2.4.12, the following #-expressions are syntactically correct:

Accurate Expression	Result Type
#> (**for** i:=1 **to** 10 **sum** (M $*$ M))	rmatrix
#< (cM + b $*$ cM + **for** i:=1 **to** 10 **sum** (cM $*$ M))	cmatrix
## (M + a $*$ M + iM $*$ iM)	imatrix
## (cM + M $*$ iM + ca $*$ iM + cM $*$ cM)	cimatrix

2.4.4.4 List of the Operands in #-Expressions

```
────────────────────────────────────── PASCAL–XSC ──
```

The definition of the functions mentioned in the following list and their declarations are given in chapter 3 and in a short form in the appendix.

real operand:
    integer variable
    integer constant
    real variable
    real constant
    inf (*interval* operand)   {lower bound of the interval}
    sup (*interval* operand)   {upper bound of the interval}
    re (*complex* operand)   {real part }
    im (*complex* operand)   {imaginary part}

interval operand:
    interval variable
    intval (*real* operand)   {transfer function}
    intval (*real* operand, *real* operand)   {transfer function}
    re (*complex* operand)   {real part}
    im (*cinterval* operand)   {imaginary part}

complex operand:
    complex variable $\{z = x + iy\}$
    conj (*complex* operand) $\{\text{conjugation } \overline{z} = x - iy\}$
    compl (*real* operand)   {transfer function}
    compl (*real* operand, *real* operand)   {transfer function}
    inf (*cinterval* operand)   {lower bound of the complex interval}
    sup (*cinterval* operand)   {upper bound of the complex interval}

cinterval operand:
    cinterval variable
    conj (*cinterval* operand)
    intval (*complex* operand)
    intval (*complex* operand, *complex* operand)
    intval (*real* operand, *complex* operand)
    intval (*complex* operand, *real* operand)
    compl (*interval* operand)
    compl (*interval* operand, *interval* operand)
    compl (*real* operand, *interval* operand)
    compl (*interval* operand, *real* operand)

rvector operand:
    rvector variable
    rvector (array variable)   {qualification}
    inf (*ivector* operand)

─────────────────────── *PASCAL-XSC* ───────────

rvector operand: (continued)
    sup (*ivector* operand)
    re (*cvector* operand)
    im (*cvector* operand)

ivector operand:
    ivector variable
    ivector (array variable)   {qualification}
    intval (*rvector* operand)
    intval (*rvector* operand, *rvector* operand)
    re (*civector* operand)
    im (*civector* operand)

cvector operand:
    cvector variable
    cvector (array variable)   {qualification}
    conj (*cvector* operand)
    compl (*rvector* operand)
    compl (*rvector* operand, *rvector* operand)
    inf (*civector* operand)
    sup (*civector* operand)

civector operand:
    civector variable
    civector (array variable)   {qualification}
    conj (*civector* operand)
    intval (*cvector* operand)
    intval (*cvector* operand, *cvector* operand)
    intval (*rvector* operand, *cvector* operand)
    intval (*cvector* operand, *rvector* operand)
    compl (*ivector* operand)
    compl (*ivector* operand, *ivector* operand)
    compl (*rvector* operand, *ivector* operand)
    compl (*ivector* operand, *rvector* operand)

rmatrix operand:
    rmatrix variable
    rmatrix (array variable)   {qualification}
    id (*rmatrix* operand)   {identity matrix}
    id (*rmatrix* operand, *rmatrix* operand)   {identity matrix}
    transp (*rmatrix* operand)   {transposed matrix}
    inf (*imatrix* operand)
    sup (*imatrix* operand)
    re (*cmatrix* operand)
    im (*cmatrix* operand)

─────────────────────────────────── *PASCAL–XSC* ───

imatrix operand:
    imatrix variable
    imatrix (array variable)   {qualification}
    id (*imatrix* operand)
    id (*imatrix* operand, *imatrix* operand)
    transp (*imatrix* operand)
    intval (*rmatrix* operand)
    intval (*rmatrix* operand, *rmatrix* operand)
    re (*cimatrix* operand)
    im (*cimatrix* operand)

cmatrix operand:
    cmatrix variable
    cmatrix (array variable)   {qualification}
    id (*cmatrix* operand)
    id (*cmatrix* operand, *cmatrix* operand)
    transp (*cmatrix* operand)
    herm (*cmatrix* operand)   {Hermitian matrix}
    conj (*cmatrix* operand)
    compl (*rmatrix* operand)
    compl (*rmatrix* operand, *rmatrix* operand)
    inf (*cimatrix* operand)
    sup (*cimatrix* operand)

cimatrix operand:
    cimatrix variable
    cimatrix (array variable)   {qualification}
    id (*cimatrix* operand)
    id (*cimatrix* operand, *cimatrix* operand)
    transp (*cimatrix* operand)
    herm (*cimatrix* operand)
    conj (*cimatrix* operand)
    intval (*cimatrix* operand)
    intval (*cimatrix* operand, *cimatrix* operand)
    intval (*rmatrix* operand, *cimatrix* operand)
    intval (*cimatrix* operand, *rmatrix* operand)
    compl (*imatrix* operand)
    compl (*imatrix* operand, *imatrix* operand)
    compl (*rmatrix* operand, *imatrix* operand)
    compl (*imatrix* operand, *rmatrix* operand)

2.4.4.5 Review of General #-Expressions

_____ *PASCAL–XSC* ———

The following tables give a complete review of #- expressions. By "special functions", we mean those listed in section 2.4.4.4.

Real and Complex Accurate Expressions

Syntax: #-Symbol (Exact Expression)

#-Symbol	Result Type	Summands Permitted in the Exact Expression
#	dotprecision	• variables, constants, and special function calls of type *integer*, *real*, or *dotprecision* • products of type *integer* or *real* • scalar products of type *real*
#* #< #>	real	• variables, constants, and special function calls of type *integer*, *real*, or *dotprecision* • products of type *integer* or *real* • scalar products of type *real*
	complex	• variables, constants, and special function calls of type *integer*, *real*, *complex*, or *dotprecision* • products of type *integer*, *real*, or *complex* • scalar products of type *real* or *complex*
	rvector	• variables and special function calls of type *rvector* • products of type *rvector* (e.g. *rmatrix* ∗ *rvector*, *real* ∗ *rvector* etc.)
	cvector	• variables and special function calls of type *rvector* or *cvector* • products of type *rvector* or *cvector* (e.g. *cmatrix* ∗ *rvector*, *real* ∗ *cvector* etc.)
	rmatrix	• variables and special function calls of type *rmatrix* • products of type *rmatrix*
	cmatrix	• variables and special function calls of type *rmatrix* or *cmatrix* • products of type *rmatrix* or *cmatrix*

```
┌─────────────────────────────────────────────────── PASCAL–XSC ──┐
```

Real and Complex Interval Accurate Expressions

Syntax: ## (Exact Expression)

#-Symbol	Result Type	Summands Permitted in the Exact Expression
##	interval	• variables, constants, and special function calls of type *integer, real, interval,* or *dotprecision* • products of type *integer, real,* or *interval* • scalar products of type *real* or *interval*
	cinterval	• variables, constants, and special function calls of type *integer, real, complex, interval, cinterval,* or *dotprecision* • products of type *integer, real, complex, interval,* or *cinterval* • scalar products of type *real, complex, interval,* or *cinterval*
	ivector	• variables and special function calls of type *rvector* or *ivector* • products of type *rvector* or *ivector*
	civector	• variables and special function calls of type *rvector, cvector, ivector,* or *civector* • products of type *rvector, cvector, ivector,* or *civector*
	imatrix	• variables and special function calls of type *rmatrix* or *imatrix* • products of type *rmatrix* or *imatrix*
	cimatrix	• variables and special function calls of type *rmatrix, cmatrix, imatrix,* or *cimatrix* • products of type *rmatrix, cmatrix, imatrix,* or *cimatrix*

2.5   Statements

In PASCAL, we distinguish between simple and structured statements. Simple statements are the assignment statement, the input/output statement, the empty statement, the procedure statement, and the **goto**-statement. Structured statements are the compound statements, the conditional statements, the repetitive statements, and the **with**-statement.

2.5.1   Assignment Statement

An assignment statement assigns the value of an expression to a variable:

| Variable := Expression

The type of the expression on the right-hand side of the assignment operator must be assignment compatible with the variable on the left-hand side (see section 2.3.5). The expression is first evaluated, and this value is assigned to the variable, i.e. the value is stored into the memory location referenced on the left-hand side. The order of access to the variable on the left-hand side and the evaluation of the expression on the right-hand side of the statement depends upon the implementation.

Within a function, the resulting value must be assigned to the function name. The function name is used like a variable of the result type on the left-hand side of the assignment statement.

Example 2.5.1:

```
var
   r, x : real;
   i, k : integer;
...
i := k div 3 + 1;
r := i div k;
r := x * x + sin (x);

i := r * x; { !! not allowed !! }
```

─────────────────────────────────────── *PASCAL–XSC* ───────

In PASCAL–XSC, a program may overload the assignment operator to assign a value to a variable of noncompatible type. This assignment overloading is done by programming the corresponding algorithm within a subroutine (see section 2.7.12).

2.5.2   Input/Output Statements

The input and output statements *read, readln, write*, and *writeln* use data files for input and output. These statements handle general files of type **file of** ..., text files of type *text*, and the standard files *input* and *output*. If no file name is specified in the corresponding statement, the standard files *input* (for reading) and *output* (for writing) are used. In this case, *input* and/or *output* must be defined in the program header as program parameters (see section 2.6). PASCAL files corresponding to external files must be listed in the program parameter list.

File Opening

The standard text files *input* and *output* are automatically opened when needed. All other files must be explicitly opened.

reset (t)	The file *t* is opened for reading. After *reset(t)*, *t*↑ contains the first element of the file. If this does not exist, then *eof(t)* is set to *true*. After the program starts, a *reset(input)* is automatically executed, *eoln(input)* is set to *true*, and *input*↑ contains a blank.
rewrite (t)	The file *t* is opened for writing. *t*↑ represents the first actual position to which data can be written.

Input Statements

read (t, v1, ..., vn)	The values for the variables *v1, ..., vn* are entered in this order from the file *t*. This statement corresponds to the statements

read (t, v1); ... read (t, vn);

Every *read (t, v)* for the general file type (*file of* ...) is defined as

begin v := t↑; get (t); **end**;

This applies also to text files (*text*), when *v* is a variable of type *char*. However, for *integer* or *real* variables, a sequence of characters is entered from the text file. This sequence must correspond to the syntax of literal constants described in section 2.3.1, and it is converted into a number. Leading blanks or end-of-line characters are ignored. The reading ends when *t*↑ cannot be a part of the number to be read (see section 2.9).

readln (t) or *readln*	The remaining characters of the current line are read, and the buffer is set to the beginning of the next line (for text files only!). The procedure *readln (t)* is defined as:

```
                    begin
                        while not eoln (t) do
                            get (t);
                        get (t);
                    end;
```

readln (t, v1, ..., vn)   corresponds to the compound statement

```
                    begin
                        read (t, v1, ..., vn);
                        readln (t);
                    end;
```

Output Statements

write (t, e1, ..., en)   The values of the expressions *e1*, *e2*, ..., *en* are evaluated and written to the file *t* in this order. This statement is equivalent to

write (t, e1); ... write (t, en);

Every *write (t, e)* for the general file type (*file of* ...) is defined as

begin t↑:= e; put (t); end;

This applies also to text files (*text*) when *e* is an expression of type *char*. For an *integer*, *real*, or *boolean* expression, a sequence of characters that represents the corresponding value is written to the text file in a standard format arranged in the necessary number of lines.

An *integer* value is represented as a decimal number without leading zeros. A sign is given only for negative values. A *real* value is represented as a decimal floating point number with one significant digit in front of the decimal point and leading minus for a negative value or blank for a positive value, and an exponential part with leading character *E*.

The logical values are written as *true* or *false*. For *char* values, the character itself (without single quotes) is written. For a character string, the sequence of characters in the string is written using the necessary number of positions.

writeln (t) or *writeln*   The current line is terminated. The next output starts from beginning of the following line (for text files only!). The procedure *writeln (t)* is defined as

> **begin**
> > t↑ := "end-of-line character";
> > put (t);
> **end**;

writeln (t, e1, ..., en)   corresponds to the compound statement

> **begin**
> > write (t, e1, ..., en);
> > writeln(t);
> **end**;

page (t)   All successive output is put on a new page (for text files only!).

Format Specifications

The form in which an *integer* expression e is printed to a text file by *write* or *writeln* can be controlled by a control expression $w > 0$ following the output parameter e in the form

> write (e : w);

The value of the *integer* expression w is called the minimum *field width* and indicates the number of characters to be written. In general, w characters are used to write e (with preceding blanks if necessary).

For output parameters of type *real*, the programmer can specify a minimum field width $w > 0$ and a fractional length $f > 0$:

> write (e : w : f);

The value of the *integer* expression f determines the digits in the fractional part (after the decimal point).

PASCAL–XSC

The procedures *reset* and *rewrite* can be called with a second parameter s of type *string*

> reset (t, s)
> rewrite (t, s)

which assigns the external (physical) file name s to the file variable t.

─────────────────────────────── *PASCAL–XSC* ───

The overloading principle available in PASCAL–XSC (see section 2.7.10) applies also to the procedures *read* and *write*. They may be overloaded to allow calling with an arbitrary number of parameters and format controls for built-in types or for user defined types (see section 2.7.11).

For the input and output of *real* values, PASCAL–XSC provides the procedures *read* and *write* (or *readln* and *writeln*) with an additional format control parameter r. This *integer* parameter specifies the rounding of the *real* value during the input or output process.

Sometimes, the value of a variable v of type *real* is entered in a form which is not exactly representable in the internal representation. The use of the statement

> read (v : r)

rounds the quantity entered according to the value of the rounding parameter r into the internal *real* format. The statement

> write (e : w : f : r)

causes the value of a *real* expression e to be rounded to f fractional places during output. For both reading and writing, the parameter r has the following meanings:

r	Rounding Mode
none	to the nearest representable number
0	to the nearest representable number
< 0	to the next-smaller representable number
> 0	to the next-larger representable number

A rounding parameter can also be used for the conversion of *real* values into strings (see section 2.9).

In order to make it possible to use the floating point output format in connection with a rounding parameter, $f = 0$ may be used as the second format control parameter. Furthermore, $w = 0$ indicates that the default floating point output format should be used.

The rounding parameter should be used for the output of values that were computed by directed-rounding operators to reflect the implementation-dependent conversion into the decimal output format.

```
                                              ——— PASCAL–XSC ———
```

Example 2.5.2:

```
    var x: real;
    begin
        read (x : +1);
        writeln (x : 11 : 0 : −1, ' ', x : 9 : 3 : 1);
    end.
```

Input: ␣4730281356200104E-12
Value of x: 4.7302813562002E3   { mantissa length 14 }
Output: ␣4.7302E+03␣␣4730.282

For the numeric types *interval, complex, cinterval, rvector, ivector, cvector, civector, rmatrix, imatrix, cmatrix,* and *cimatrix,* the overloading of *read* and *write* is predefined in the arithmetic modules (see chapter 3). For new user-defined data types, *read* and *write* can be overloaded by explicit declarations (see section 2.7.11).

2.5.3   Empty Statement

The empty statement can be used at places where syntactically a statement is necessary, but no action is intended by the programmer. There is no special symbol for the empty statement. It is recognized from context, for example, between two symbols

    ; ;   or   ; **end**   or   **then else**   etc.

The empty statement is meaningfully used in connection with the **goto**-statement when branching to the end of a block.

Example 2.5.3:

```
    goto 100;
        ⋮
    100:   { empty statement }
    end;
```

In this book, we include an empty statement before each **end** so that statements can be added to the end of a block without requiring the programmer to add a ; to the end of the existing code.

2.5.4   Procedure Statement

A procedure statement causes the call of the named procedure with the actual parameters replacing the formal parameters:

ProcedureIdentifier
> (ActualParameterList)   { may be omitted }

If the procedure is declared without formal parameters, then the procedure must be called without an actual parameter list. Otherwise, the actual parameters must be consistent with the formal parameters in the same order. With a call by reference, the actual parameter must be a variable of compatible type. With a call by value, the actual parameter must be an expression that is assignment compatible to the formal parameter. Further details are found in 2.7.1.

PASCAL–XSC

PASCAL–XSC allows a modified call by reference in connection with structured data types (see section 2.7.9).

Example 2.5.4:

> quicksort (x, i, j);   { call of a sort procedure }
> primenumber(m);   { call of a prime number generating procedure }

2.5.5   goto-Statement

The **goto**-statement indicates that further processing should continue at another part of the program. The sequential execution of the program is broken, and processing is continued at a labeled statement.

All statement labels must be declared in the declaration part of the corresponding block. The declaration is:

label
> LabelList; { not empty }

The **goto**-statement has the form:

goto Label

The label is an unsigned integer with a maximum of four digits. The labeled statement has the form:

Label : Statement

A **goto**-statement may only branch to a label that marks a statement of the same or a higher level according to the block structure of the program.

The **goto**-statement should be used with caution!

2.5.6 Compound Statement

A compound statement combines a sequence of statements into a single statement:

> **begin**
>   Statement; ...
> **end**

The execution of a compound statement is analogous to the execution of the statement part of a program.

Example 2.5.5:

> while i <= n do
> begin
>   s := s + a[i];
>   i := i + 1;
> end;

2.5.7 Conditional Statements

2.5.7.1 if-Statement

The if-statement allows the selective execution of two statements:

> **if** LogicalExpression **then** Statement
>   **else** Statement { may be omitted }

The execution of the if-statement causes the evaluation of the logical expression. If the value of the expression is *true*, the statement after **then** (1st alternative) is executed. Otherwise the statement after **else** (2nd alternative) is executed. The **else** alternative may be omitted. This situation is handled as if the **else** alternative were an empty statement.

Example 2.5.6:

> if x <= y   then   z := y − x
>   else   z := x − y;   { positive difference of x and y }
> if x >= 0   then   y := sqrt (x);

In nested if-statements, the rule applies that every **else** goes with the closest **if**.

2.5.7.2 case-Statement

While the if-statement handles only two alternatives, the **case**-statement allows the execution of a statement which is chosen from arbitrarily many alternatives:

> **case** IndexExpression **of**
>   ConstantList: Statement; ...   { not empty }
> **end**

The first operation of the **case** statement is the evaluation of the index expression. If the value of this expression is contained in one of the constant lists, the corresponding statement is executed. If the value of the expression is not in a constant list, an error message is given.

The index expression may be of type *integer*, *boolean*, *char*, or an enumeration type. The constants in all of the constant lists must be of the same type. Successive constants of a constant list may be abbreviated in the form of a subrange according to the ordering of the basic types.

| Constant .. Constant

All constant lists must be disjoint.

PASCAL–XSC ──

The **case** statement may contain an **else**-alternative immediately before **end**. This alternative covers all constants which are not listed in the constant lists of the **case**-statement:

| **else**: Statement

This **else** alternative is executed when the value of the index expression is not listed in one of the constant lists.

Example 2.5.7:

```
case trunc (phi/90) + 1 of
    1: f := phi * r;
    2: f := 90 * r;
    3: f := -(phi - 180) * r;
    4: f := - 90 * r;
    else f := 0;
end;
```

2.5.8   Repetitive Statements

2.5.8.1   while-Statement

The **while**-statement allows the repetitive execution of a statement under the control of a beginning condition:

| **while** LogicalExpression **do** Statement

The statement following **do** is executed as long as the logical expression has the value *true*. Hence, the logical expression is evaluated before each execution of the statement. If the value of the expression is *false*, the **while**-statement is terminated. This can happen before the execution of Statement for the very first time.

Example 2.5.8:

```
i := n;
while i >= 1 do
begin
    s := s + a[i];
    i := i - 2;
end;
```

2.5.8.2   repeat-Statement

The **repeat**-statement executes a series of statements until an end condition is fulfilled.

> **repeat**
>   Statement; ...
> **until** LogicalExpression

The statements between **repeat** and **until** are executed repeatedly until the logical expression evaluates to *true*. The logical expression is evaluated *after* every execution of the series of statements. This means that the sequence is executed at least once.

Example 2.5.9:

```
i := n;
repeat
    s := s + a[i];
    i := i - 2;
until i < 1
```

The two statements between **repeat** and **until** are executed at least once, no matter what the value of *n*. For $n = 0$, the statements *s:= s + a[0];* and *i:= −2* are executed.

2.5.8.3   for-Statement

The **for**-statement allows the repetitive execution of a statement for a known number of repetitions:

> **for** ControlVariable := InitialValue **to** FinalValue **do**
>   Statement

or

> **for** ControlVariable := InitialValue **downto** FinalValue **do**
>   Statement

The first action of the **for**-statement is to evaluate the expressions for the initial and final value. If the final value is smaller (or, in the case of **downto**, larger) than the initial value, the execution of the **for** statement is ended. This situation is referred to as an *empty loop*. Otherwise, the control variable is set to the starting value and the statement is executed. If the control variable is not equal to the final value, it is incremented (or, in the case of **downto**, decremented), and the statement is executed repeatedly until the final value is reached. The control variable may be of types *integer, boolean, char,* or enumeration type and must be declared in the same block as the **for**-statement. The initial and final values must be of compatible types.

The expressions for the initial and the final values are only evaluated once at the beginning of the execution of the **for**-statement. However, it is considered poor programming practice to change the initial or the final value within the loop.

Within the statement after the **do**, the control variable *may not* occur

- on the left-hand side of an assignment statement,

- as an actual parameter for a formal **var**-parameter of a subroutine call,

- as an input parameter of a *read* statement, or

- as a control variable in a further **for**-statement.

On exit from the **for**-statement, the value of the control variable is considered undefined.

Example 2.5.10:

```
for i := 1 to n do s := s + a[i];
for i := n downto 1 do s := s + a[i];
```

A variation of the step length can only be accomplished through additional and explicit programming.

Example 2.5.11:

```
for i := 1 to n do s := s + a[2*i];
```

2.5.9   with-Statement

The **with**-statement facilitates working with records by allowing an abbreviated notation for the record components. The **with**-statement has the form

❚   with RecordVariableList **do** Statement

The list can contain more than one variable after the reserved word **with**, for example:

```
with r1, r2, ..., rn do statement;
```

This corresponds to the nesting of the **with**-statements

   with r1 **do with** r2, ..., rn **do** statement;

So, it suffices to explain the execution of the **with**-statement

   with r **do** statement;

which is equivalent to the execution of the statement after the **do** using the record components of r. The advantage of the **with**-statement is that the components of r can appear without the prefix r. in this statement.

Example 2.5.12:

```
type date = record day: 1..31;
                    month: (Jan, Feb, Mar, Apr, May, Jun,
                            Jul, Aug, Sep, Oct, Nov, Dec);
                    year: integer;
            end;
var birthday: date;
  ...
begin
  ...
  with birthday do
  begin
     day := 4;
     month := Dec;
     year := 1960;
  end;
  ...
end.
```

2.6   Program Structure

A program consists of a program header, a declaration part, and a statement part (body) between **begin** and **end**. The program header contains the name of the program that is specified after the reserved word **program**, and optionally the program parameters, i.e. the names of the external files used (especially *input* and *output*).

> **program** Name
>   (ProgramParameterList)   { may be omitted }
>   ;
>   Declaration; ...
> **begin**
>   Statement; ...
> **end**.

The statement part describes the processing steps (algorithm) which are executed by the computer. All objects appearing in this part that are not predefined standard objects must be defined in the declaration part. In Standard PASCAL, the order of the declaration sections is: label declaration part, constant declaration part, type declaration part, variable declaration part, and finally the procedure and function declaration part.

When coding the program, note the following rules for the use of separating symbols:

- No separating symbol may occur within a name, number, reserved word, or a two-character symbol (e.g. <=, :=).

- Identifiers, numbers, or reserved words immediately following one another must be separated by at least one separating symbol.

The separating symbols are the blank space (⊔), a tab character, a new line, or a comment, which appears within braces "{", "}".

The execution of the program causes the processing of the declarations in the given order. Then, the execution of the statements begins with the physically first statement. After each statement, the following statement is executed. Normally, this is the physically next statement, but this is not necessarily the case in **goto** and structured statements.

───────────────────────────────────── *PASCAL–XSC* ───────

An executable program consists of a main program, as in Standard PASCAL, and possibly of a number of modules, which are introduced by a **use**-clause in the main program itself or in a used module.

In the main program, objects can appear that are

- predefined,
- defined or declared in the main program, or
- globally defined in used modules.

A main program has the form:

> **program** Name
>   (ProgramParameterList)   { may be omitted }
>   ;
>   UseClause; ...
>   Declaration; ...
> **begin**
>   Statement; ...
> **end.**

A program first executes the used modules. Then, the processing continues with the execution of the declarations in the main program and finally, the execution of the statements, as in Standard PASCAL.

The declarations preceded by **label, const, type, var, function, proce-dure, priority,** or **operator** may appear more than once and in any order. An identifier must be declared or defined before it is used (except see pointers, section 2.3.4).

2.7   Subroutines

Special parts of algorithms in PASCAL may be declared and called as *procedures* or *functions*. The purpose of a function is to execute an algorithm and return a single result of type *integer, real, boolean, char*, an enumerated type, or a pointer type. A procedure is an algorithm which can return any number of parameters with each of them possibly having a different type. The declaration of procedures and functions occurs immediately before the statement part of a program.

─── *PASCAL–XSC* ───

As a further extension to subroutines, PASCAL–XSC has the option to declare *operators* whose result, like that of a function, can be of any type. Procedures, functions, and operators can be declared anywhere within the declaration part of a program.

2.7.1   Procedures

The form of a procedure declaration is very similar to that of a program:

> **procedure** Name
>   (FormalParameterList) { may be omitted }
>   ;
>   Declaration; ...
> **begin**
>   Statement; ...
> **end;**

The formal parameter list describes those objects of the procedure which serve as input and output parameters. Formal parameters can be variables, procedures, or functions. The specification of parameters has the form:

> **var**   { may be omitted }
>   IdentifierList: TypeSpecification

If the reserved word **var** precedes an identifier or an identifier list, then the listed variables are used for a call by reference (variable parameters) when the procedure is called. Otherwise, the variables are used for a call by value (value parameters).

The specification of procedures and functions is given by a corresponding procedure or function header along with the formal parameters and the type of the function. The sections of the formal parameter list are each separated with a semicolon (;). There are no limits to the length and the order of the list.

In contrast to declarations, the type specification for formal parameters may contain a conformant array scheme:

| **array** [IndexRangeList] **of** TypeSpecification

with index ranges of the form

| Identifier..Identifier : Type

and the separating symbol ; in the index range list. A conformant array scheme leaves the index bounds of the formal argument indeterminate until the procedure is called. The identifiers that are specified in this scheme can be used to access to the index bounds inside the procedure.

The statement part of the procedure contains the program statements that implement the algorithm. These statements can be formulated using

- formal parameters,

- local objects of the procedure (i.e. objects declared within the procedure), and

- non-local objects of the procedure (i.e. objects of the encompassing program or procedure).

Example 2.7.1:

```
type
    fraction = record N, D : integer end;

procedure readfraction (var b: fraction);
    begin
        write ('Numerator = '); read (b.N);
        write ('Denominator = '); read (b.D);
    end;

procedure addfraction (a, b: fraction; var g: fraction);
    begin
        g.N:= a.N * b.D + b.N * a.D;
        g.D:= a.D * b.D;
    end;
```

The call of a procedure is given by a procedure statement:

| ProcedureIdentifier
                        (ActualParameterList) { may be omitted }

A procedure statement handles the parameter list in the manner described below and then executes the statement part of the called procedure.

- The actual parameters are related to the formal parameters in the given order. With a call by reference, the rules of type compatibility are applied. With a call by value, the rules of assignment compatibility are applied (see 2.3.5).

- The formal parameters representing variables for a call by reference are used to access the corresponding actual parameters during the execution of the procedure.

- For the formal parameters representing variables for a call by value, memory is allocated, and the values of the actual parameters (expressions) are assigned to them before the statement part of the procedure is executed.

- During the execution of the procedure, formal procedure and function parameters serve as names for the corresponding actual procedures and functions.

PASCAL–XSC

PASCAL–XSC allows a modified call by reference in connection with structured data types (see 2.7.9).

Example 2.7.2:

```
var
    a, b, g : fraction;
begin
    readfraction (a);
    readfraction (b);
    addfraction (a, b, g);
    ...
end.
```

In the statement part of a procedure, local and non-local subroutines may be called. A procedure may call itself (*recursion*). This recursive call may occur directly or indirectly. Fundamentally, the called procedure must be declared before it is called. This declaration can be accomplished incompletely by the use of a **forward** declaration (see section 2.7.8).

PASCAL–XSC

Instead of the conformant array scheme, the more powerful concept of dynamic arrays is available (see section 2.3.2). Through the use of a dynamic type for a formal parameter, the index range remains indeterminate until the procedure is called with actual parameters. The access to the index bounds within the procedure body is managed through the use of the functions *lbound* and *ubound*.

—————————————————————————————— *PASCAL–XSC* ——

Example 2.7.3:

> **type** dynvector = **dynamic array** [*] **of** real;
> . . .
> **procedure** vecadd (**var** x, y, res: dynvector);
>   { equal index bounds for x, y, res are assumed }
>   **var** i: integer;
>   **begin**
>   **for** i:= lbound(x) to ubound(x) **do**
>   res[i] := x[i] + y[i]
>   **end**;

The call of the procedure *vecadd* can only occur with vectors of type *dyn-vector*. Using this implementation, the index ranges of the actual parameters x and y must match with the index range of the actual parameter *res*.

In this example, x and y are specified as var parameters to save the storage which would be required for copying a call-by-value parameter (see section 2.7.9).

If a function which returns a result of a dynamic type appears as a formal parameter in a procedure, then the function header may only contain the name of the dynamic type without the index bounds.

2.7.2   List of Predefined Procedures and Input/Output Statements

The following predefined procedures and input/output statements are available in Standard PASCAL:

Allocation and Release of Referenced Variables:

> new (PointerVariable)
> new (PointerVariable, TagFieldValue, ..., TagFieldValue)
> dispose (PointerVariable)
> dispose (PointerVariable, TagFieldValue, ..., TagFieldValue)

Reading and Writing on File Variables:

> reset (FileVariable)
> get (FileVariable)
> read (FileVariable, Variable, ..., Variable)
> readln (TextFileVariable, Variable, ..., Variable)
> rewrite (FileVariable)
> put (FileVariable)

write (FileVariable, Variable, ..., Variable)
writeln (TextFileVariable, Variable, ..., Variable)
page (TextFileVariable)

PASCAL–XSC

The PASCAL–XSC extensions are:

Allocation and Release of Referenced Variables:

mark (PointerVariable)
release (PointerVariable)

Reading and Writing on File Variables:

reset (FileVariable, StringExpression)
rewrite (FileVariable, StringExpression)

Changing the Actual Length of String Variables:

setlength (StringVariable, IntegerExpression)

2.7.3   Functions

A partial algorithm that delivers only one result of a simple type (*integer*, *real*, *boolean*, *char*, an enumeration type, or a pointer type) can be formulated as a function in place of a procedure:

function Identifier
   (FormalParameterList)   { may be omitted }
   : Type;
Declaration; ···
begin
   Statement; ···
end;

The result of a function is returned by the name of the function, and not by a formal parameter. The type of the function (or of the function result) is specified following the formal parameter list after the colon (:). The function value must be assigned to the name of the function in the statement part of the function. Thus, the function name may appear on the left-hand side of the assignment statement.

The appearance of the function name on the right hand side of an assignment statement is a recursive call of the function. All other rules for the declaration of functions are analogous to those for procedures.

The calling of a function has the form

> FunctionIdentifier
> > (ActualParameterList) { may be omitted }

and serves as an operand in an expression. The evaluation of the expression is interrupted, parameters are handled as described for procedures on page 86, the statement part of the function is executed, and the value computed within the function is assigned to the function identifier. Then, the evaluation of the expression is continued by using the function result in place of the function call.

PASCAL–XSC

PASCAL–XSC allows a modified call by reference in connection with structured data types (see section 2.7.9).

In the statement part of a function, local and non-local subroutines may be called. The use of the function itself is a *recursive* execution of the function. The recursive call can occur either as a direct call or as an indirect call from another function. Fundamentally, the called function must be declared before it is called. This declaration can be accomplished incompletely by the use of a **forward** declaration (see section 2.7.8).

2.7.4   Functions with Arbitrary Result Type

PASCAL–XSC

PASCAL–XSC removes the restriction of function result types to *integer, real, boolean, char*, an enumeration type, or a pointer type. A function result may be of any structured type. The assignment to the function result may be done componentwise or by assigning the entire structure as a unit. For a record type, the use of the **with**-statement for the function result is also possible.

Example 2.7.4:

        type mycomplex = **record** re, im : real **end**;
        . . .
        function mycompladd (y, w: mycomplex) : mycomplex;
          begin
              mycompladd.re := y.re + w.re;
              mycompladd.im := y.im + w.im;
          end;

Furthermore, a dynamic type can be used for the function result. The index bounds of the dynamic result are specified by expressions that must be able to be evaluated before the execution of the function body.

```
                                            ──── PASCAL-XSC ────

Example 2.7.5:
    type dynvector = dynamic array [*] of real;
    function vecadd (x, y: dynvector) :
                dynvector [lbound(x)..ubound(x)]; {function type}
    { the same index bounds for x and y are assumed }
      var i: integer;
      begin
        for i:= lbound(x) to ubound(x) do
            vecadd[i] := x[i] + y[i];
      end;

If a function with a dynamic result appears as a formal parameter of a procedure,
then the function header may only contain the name of the dynamic type without
the index bounds.
```

2.7.5   List of Predefined Functions

Here are the predefined functions of Standard PASCAL, grouped according to the
allowed parameter types. The types of the function results are given in braces.

Parameter type *integer, boolean, char,* enumeration type

ord (Expression)	{ *integer* }
succ (Expression)	{ Parameter type }
pred (Expression)	{ Parameter type }

Parameter type *integer*

odd (Expression)	{ *boolean* }
chr (Expression)	{ *char* }

Parameter type *integer, real*

abs (Expression)	{ Parameter type }
sqr (Expression)	{ Parameter type }
sqrt (Expression)	{ *real* }
exp (Expression)	{ *real* }
ln (Expression)	{ *real* }
arctan (Expression)	{ *real* }
sin (Expression)	{ *real* }
cos (Expression)	{ *real* }
round (Expression)	{ *integer* }
trunc (Expression)	{ *integer* }

Parameter type File

eof (FileVariable) or eof	{ *boolean* }
eoln (TextFileVariable) or eoln	{ *boolean* }

──────────────────────────────── PASCAL–XSC ────

The PASCAL–XSC extensions are:

Arbitrary parameter type
    loc (Variable)   *{ integer }*

Parameter type Pointer
    ord (P Expression)   *{ integer }*

Parameter type *integer, real, dotprecision*
    sign (Expression)   *{ integer }*

Parameter type *real*   *{ Result type real }*
    succ (R Expression)
    pred (R Expression)
    exp2 (R Expression)
    exp10 (R Expression)
    log2 (R Expression)
    log10 (R Expression)
    tan (R Expression)
    cot (R Expression)
    arcsin (R Expression)
    arccos (R Expression)
    arccot (R Expression)
    arctan2 (R Expression, R Expression)
    sinh (R Expression)
    cosh (R Expression)
    tanh (R Expression)
    coth (R Expression)
    arsinh (R Expression)
    arcosh (R Expression)
    artanh (R Expression)
    arcoth (R Expression)

Parameter type Array   *{ Result type: Array index type }*
    lbound (ArrayVariable, I Constant)
    lb (ArrayVariable, I Constant)
    lbound (ArrayVariable)
    lb (ArrayVariable)
    ubound (ArrayVariable, I Constant)
    ub (ArrayVariable, I Constant)
    ubound (ArrayVariable)
    ub (ArrayVariable)

─────────────────────── *PASCAL–XSC* ───

Parameter Type *integer, real*   { Result type *string* }
    image (I Expression)
    image (I Expression, I Expression)
    image (R Expression)
    image (R Expression, I Expression)
    image (R Expression, I Expression, I Expression)
    image (R Expression, I Expression, I Expression, I Expression)

Parameter Type *string*
    ival (ST Expression)   { *integer* }
    ival (ST Expression, ST Variable)   { *integer* }
    rval (ST Expression)   { *real* }
    rval (ST Expression, ST Variable)   { *real* }
    rval (ST Expression, I Expression)   { *real* }
    rval (ST Expression, I Expression, ST Variable) { *real* }
    length (ST Expression)   { *integer* }
    maxlength (ST Variable)   { *integer* }
    pos (ST Expression, ST Expression)   { *integer* }
    substring (ST Expression, I Expression, I Expression) { *string* }

Additional predefined functions for the data types *complex, interval, cinterval, rvector, cvector, ivector, civector, rmatrix, cmatrix, imatrix,* and *cimatrix* are provided in the modules C_ARI, I_ARI, CI_ARI, MV_ARI, MVC_ARI, MVI_ARI, and MVCI_ARI (see chapter 3).

2.7.6   Operators

─────────────────────── *PASCAL–XSC* ───

PASCAL–XSC lets the programmer define subroutines in the form of operator declarations. We have two different kinds of operators, i.e. operators *with result* and operators *without result*. The assignment operator := is the only operator without result. Its definition and overloading is described in detail in section 2.7.12.

A programmer may define unary and binary operators with arbitrary operand type and arbitrary result type. User-defined operators may be used in expressions interchangeably with built-in operators. User-defined operators are declared in a form similar to a function declaration.

PASCAL–XSC

The declaration is

> **operator** MonadicOperator (FormalParameter)
>   ResultIdentifier: TypeSpecification;
> Declaration; ...
> **begin**
>   Statement; ...
> **end**;

or

> **operator** DyadicOperator (FormalParameter, FormalParameter)
>   ResultIdentifier: TypeSpecification;
> Declaration; ...
> **begin**
>   Statement; ...
> **end**;

Thus, unary operators have exactly one operand, and binary operators have exactly two operands. The result identifier takes the place of the function identifier. The assignment to the result must occur in the operator body. The formal parameters are listed in the form

> **var** { may be omitted }
>   Identifier: TypeSpecification

If both operands have the same type and both are either reference or value parameters, the specification may be shortened to

> **var** { may be omitted }
>   Identifier, Identifier: TypeSpecification

──────────────────────────────── *PASCAL-XSC* ────

The programmer may overload the names of the monadic operators

$+, -,$ **not**   (priority 3)

and the dyadic operators

$=, <>, <=, >=, <, >,$ **in**, $><$ (priority 0)
$+, +<, +>, -, ->, -<, +*,$ **or**   (priority 1)
$*, *<, *>, /, /<, />, **,$ **mod, div, and**   (priority 2)

Furthermore, an

Identifier

defined by the user may be introduced as a new operator symbol. A new operator identifier must first occur in a priority definition:

| **priority** Identifier = PrioritySymbol; ...

This definition fixes the priority of the new identifier corresponding to the symbols $=, +, *,$ and $\uparrow$. The priority symbols $=, +,$ and $*$ correspond to binary operators with priority 0 ($=$), 1 ($+$) and 2($*$), whereas the symbol $\uparrow$ corresponds to a monadic operator with priority 3.

We speak of *overloading* of an already existing operator if the declaration is given with alternate operand types. Hence, various overloaded operators can be distinguished by their operands. For example in Standard PASCAL, the operator $+$ is already overloaded (on the one hand for *integer* addition, on the other hand for *real* addition). In Example 2.7.6, we overload the operator $+$ to provide addition of vectors. If a given operator becomes redefined with the same operand types (*concealment*), the existing operator is hidden. In this case, the operator symbol has a new meaning for the same operand types for which it was previously defined. The old meaning is hidden according to the underlying block structure (see section 2.7.10).

Example 2.7.6:

```
type
    mycomplex = record re, im : real end;
    dynvector = dynamic array [*] of real;
operator * (z, w: mycomplex) complmult: mycomplex;
    begin
        complmult.re := z.re * w.re - z.im * w.im;
        complmult.im := z.re * w.im + z.im * w.re;
    end;
```

─────────────────────────────────────── *PASCAL–XSC* ───

operator + (x, y: dynvector) vecadd: dynvector [lb(x)..ub(x)];
  { the same index bounds for x and y are assumed }
  var i: integer;
  begin
    for i:= lbound(x) **to** ubound(x) **do**
      vecadd[i] := x[i] + y[i];
  end;
priority xor = +; { exclusive or }
operator xor (a, b: boolean) exor: boolean;
  begin
    exor := a <> b;
  end;

A monadic operator is used or "called" within an expression by

| OperatorSymbol ActualParameter

a dyadic operator by

| ActualParameter OperatorSymbol ActualParameter

Like functions, operators can only occur within expressions. The evaluation of the expression containing the operator is interrupted, the actual parameters are handled as described in connetion with procedures on page 86, and the statement part of the operator is executed. Finally, the evaluation of the expression is continued using the result of the operator instead of the operator call.

PASCAL–XSC permits a modified call by reference in connection with structured types (see section 2.7.9).

Example 2.7.7:

    var
        complex_1, complex_2, complex_3, complex_4 : mycomplex;
        vector_x, vector_y, vector_z : dynvector [1..100];
        boolean_1, boolean_2, boolean_3 : boolean;
    begin
        complex_4 := complex_1 * complex_2 * complex_3;
        vector_z := vector_x + vector_y;
        boolean_3 := boolean_1 xor boolean_2;
    end.

The statement part of an operator may contain calls to local or non-local subroutines, including the operator itself. This recursive call may be affected either directly or indirectly. An operator always must be declared before it is used. This declaration may also be done incompletely in form of a **forward**-declaration (see section 2.7.8).

2.7.7   Table of Predefined Operators

-- *PASCAL–XSC* ----------

The predefined operators of PASCAL and the extensions of PASCAL–XSC are listed in the following table. The additional operators provided in the arithmetic modules for the arithmetic types *complex, interval, cinterval, rvector, cvector, ivector, civector, rmatrix, cmatrix, imatrix,* and *cimatrix* are not listed here. The corresponding tables are given in chapter 3 and in appendix B.4.

left Operand \ right Operand	integer	real	boolean	char	string	set
monadic	+, −	+, −	not	'		
integer	○, ○<, ○>, div, mod, ∨	○, ○<, ○>, ∨				in
real	○, ○<, ○>, ∨	○, ○<, ○>, ∨				
boolean			or, and, =, <>, <=, >=	'		in
char				+ ∨	+ ∨ in	in
string				+ ∨	+ ∨ in	
set						+, −, *, =, <>, <=, >=
enumeration type						in

Predefined Operators of PASCAL–XSC

$$\circ \in \{+, -, *, /\}$$
$$\vee \in \{=, <>, <, <=, >, >=\}$$

2.7.8   forward- and external-Declaration

─────────────────────────────────── *PASCAL–XSC* ───────

The **forward**-declaration allows routines to be called mutually or recursively.
This *incomplete declaration* of procedures, functions, or operators is given by
the head of the procedure, function, or operator followed by the reserved word
forward instead of the body of the procedure, function, or operator. The complete declaration of such a subroutine must occur in the same declaration part.
This complete declaration is also introduced with the reserved word **procedure**,
function, or **operator** and the corresponding identifiers. In contrast to Standard PASCAL, the formal parameter list, the result identifier (for operators),
and the result type specification (for functions and operators) must be listed
once again.

An **external**-declaration allows separately compiled procedures, functions,
or operators written in a different language or in assembler to be linked. The
reserved word **external** appears instead of the declaration part and body of the
routine. Optionally, a string constant may follow. The identifier of the external
subroutine is either the identifier of the procedure or function, the result identifier
(for operators), or the value of the string constant following the reserved word
external. This means that external subroutines may be overloaded, because the
same internal name can be used for different external routines. The specification
of the formal parameter list only serves for the syntactical control of the subroutine calls. A detailed description of the use of external subroutines in connection
with **external** is given in the implementation-dependent user manual.

2.7.9   Modified Call by Reference for Structured Types

─────────────────────────────────── *PASCAL–XSC* ───────

Usually, operators and functions are used in a nested way. Within an expression,
operators or functions are called repeatedly. Thus, expressions should be permitted as actual operands or as actual parameters. In the strict sense of PASCAL,
this means that the formal operands and formal parameters must be declared
and used as value parameters, because otherwise no expressions can take the
places of the parameters. So, the use of the operator or the call of the function
causes local memory to be allocated for the copies of the actual parameters. This
is very inefficient with large structured types.

—————————————————————————————— *PASCAL–XSC* ——————

To avoid this, PASCAL–XSC allows a modified call by reference for structured types. The actual parameters corresponding to formal **var**-parameters may be given by expressions. During execution of the routine, the formal **var**-parameter is used as an access to the anonymous auxiliary quantity allocated by the compiler during evaluation of the expression and containing the value of the expression.

Example 2.7.8:

With the declarations

```
const
    n = 100;
type
    matrix = array [1..n, 1..n] of real;
var
    m1, m2, m3, m4, m5 : matrix;
operator + (var a, b: matrix) resplus : matrix;
    var i, j: integer;
    begin
        for i:= 1 to n do
            for j:= 1 to n do
                resplus[i,j] := a[i,j] + b[i,j];
    end;
function component_sqr (var a: matrix) : matrix;
    var i, j: integer;
    begin
        for i:= 1 to n do
            for j:= 1 to n do
                component_sqr[i,j] := sqr (a[i,j]);
    end;
```

an assignment statement of the form

```
m1 := m1 + m2 + component_sqr (m3 + m4 + m5);
```

is permitted.

2.7.10   Overloading of Procedures, Functions, and Operators

─────────────────────────────── *PASCAL–XSC* ──────

PASCAL–XSC procedures, function and operators are identified by their names (symbols) and by number, type, and order of parameters. Thus, in contrast to Standard PASCAL, several procedures, functions, and operators with the same name may be defined within a block, as long as the compiler can distinguish them by their parameters. This feature is called *overloading* of the identifiers. In Standard PASCAL, an exponential function for complex numbers must be declared by the use of a name different from *exp*, which is used for the predefined *real* function. In PASCAL–XSC, however, the predefined function identifier *exp* may be overloaded for use with parameters of user-defined types.

Example 2.7.9:

>   **type** complex = **record** re, im : real **end**;
>   ...
>   **function** exp (c : complex) : complex;
>   **begin**
>   exp.re:= exp (c.re) * cos (c.im);
>   exp.im:= exp (c.re) * sin (c.im);
>   **end**;

>   The *real* function *exp* is called with the real parameter *c.re* within the body. Hence, this is not a recursive call of the newly defined function *exp*.

The following rules apply to the overloading of procedures, functions, and operators (called routines in the following):

- The formal parameter lists of overloaded routines must be different, i.e. the parameters must not agree in number, type, and order simultaneously. In this context, the difference between value- and **var**-parameter is insignificant. Compatible types are handled as the same types.

- The result type of functions and operators is not significant for the identification.

- Functions may be overloaded only by functions, operators only by operators, and procedures only by procedures.

- Within the same block, a routine identifier may not be used as identifier for a constant, a variable, or a type simultaneously.

─────────────────────────────────────── *PASCAL-XSC* ───────

The rules of concealment are the same as in Standard PASCAL. An identifier is concealed if the same identifier is declared in an inner block. Routines of the outer block are not concealed if they are overloaded in the inner block with different parameter lists. The following rules apply to the call of an overloaded subroutine:

- A call by reference requires the actual parameters to be compatible with the formal parameters.

- A call by value requires the actual parameters to be assignment compatible with the formal parameters. If no routine with parameters of compatible type is available, then the assignment compatible actual parameters may be converted automatically. This context assumes the strict interpretation of the assignment compatibility, i.e. an overloading of the assignment operator does *not* make the corresponding types assignment compatible for the automatic conversion (see section 2.3.5 and section 2.7.12).

If a routine call matches several overloaded procedures, functions, and operators, then the ambiguity is resolved as follows. If there is a routine whose formal parameters exactly match with the actual parameters of the call (concerning reference and value parameters), then this one is chosen. If this is not the case, then a routine is chosen that allows assignment of the actual value parameters to formal reference parameters (see also section 2.7.9) without conversion of conforming type. Otherwise, the routine is chosen which has the *first* parameter whose type is compatible, and not merely conforming.

Example 2.7.10:

    operator +∗ (a: integer; b: real) ir_res: real;
    . . .
    operator +∗ (a: real; b: integer) ri_res: real;
    . . .
    var
       i : integer;
       r, res : real;
    . . .
    res:= i +∗ r;   { 1st operator is used }
    res:= r +∗ i;   { 2nd operator is used }
    res:= i +∗ i;   { 1st operator is used }
    res:= r +∗ r;   { assignment not possible }

In the third assignment statement, neither the first nor the second operator matches exactly. Either one could be used by converting the *integer i* to a *real* number. According to the rule above, we choose the first operator because its first operand is an *integer*.

2.7.11   Overloading of *read* and *write*

-- *PASCAL–XSC* -------

The overloading described in the preceding section also applies to the procedures *read* and *write*. Since these procedures have some special features in Standard PASCAL, the concept of overloading has been modified for these input/output routines.

In section 2.5.2, we mentioned that *read* and *write* in connection with text files permit

- an optional first parameter of type *text*,

- an arbitrary number of different parameters, and

- optional format specifications following an input/output element separated by a colon.

By overloading of *read* and *write* in PASCAL–XSC, these features are also supported for user-defined input/output procedures. We must consider some rules for the declaration and call.

Declaration

The first parameter of a newly declared input/output procedure must be a **var**-parameter of type *text* or of any arbitrary file type. The second parameter represents the quantity to be input or output, and must not be a file type. All following parameters are interpreted as format specifications for the second parameter.

Example 2.7.11:

> **type** interval = **record** inf, sup : real **end**;
>
> . . .
>
> **procedure** write (**var** f: text; int: interval; m, n: integer);

Call

The file parameter may be omitted when calling an overloaded input/output procedure. This corresponds to a call with the standard file *input* or *output*. If a file parameter is given, the second actual parameter (otherwise the first) is the input/output object. The format parameters for this parameter follow, separated by a colon.

────── *PASCAL–XSC* ──────

Example 2.7.12:

With *int* of type *interval* and *f* of type *text*, the output procedure declared above may be called by

write (int : 10 : 5);   or   write (f, int : 12 : 6);

Several input statements or output statements can be combined to a single statement as in Standard PASCAL.

Example 2.7.13:

With a *real* variable a, the statement

writeln (f, a : 20 : 9, int : 50 : 10, true : 4);

is equivalent to the statements

write (f, a : 20 : 9);
write (f, int : 50 : 10);
write (f, true : 4);
writeln (f);

For each of these *write*-calls, the compiler is looking for a user-defined procedure with corresponding parameters interpreting every colon as a comma. If there is no such procedure available, the standard input or output procedure is used, if possible.

To supply the input or output for various number of format parameters, the user must implement a procedure for every number of format parameters (see Example 2.7.15).

Example 2.7.14:

If we do not want to specify the rounding of *real* numbers by the *integer* parameter as usual, we could implement the following procedures for example:

procedure write (**var** f: text; r: real;
                            w, n: integer; rd: boolean);
    begin
      if rd **then**
        write (f, r : w : n : +1)
      else
        write (f, r : w : n : −1);
    end;

Example 2.7.15:

Further variants of the format specification:

```
procedure write (var f: text; r: real;
                      w: integer; rd: boolean);
   begin
      write (f, r : w : 0 : rd)
   end;
   procedure write (var f: text; r: real; rd: boolean);
      begin
         write (f, r : 20 : 0 : rd)
      end;
```

With these declarations, the output of the *real* expressions a, b, c can be done by

```
writeln (output, a : 10 : 5 : true, b : 10 : false, c : true);
```

A final example demonstrates the universal applicability of overloading of *read* and *write*.

Example 2.7.16:

```
const
   format1 = '[ ]';
   format2 = '<>';
   format3 = '( )';
...
procedure write (var f: text; int: interval; parenth: string);
var l, r: char;
begin
   l:= parenth[1];
   r:= parenth[2];
   write (f, l, int.inf : 20 : 13, ',', int.sup : 20 : 13, r);
end;
```

With these declarations, intervals may be written in different forms:

```
with    write (int : format1);    in the form    [ ... , ... ]
with    write (int : format2);    in the form    < ... , ... >
with    write (int : format3);    in the form    ( ... , ... )
```

Using the possibilities of overloading, even format specifications similar to those of FORTRAN may be realized.

2.7.12   Overloading of the Assignment Operator :=

──────────────────────────── PASCAL–XSC ──────

The programmer can overload the assignment operator := as an operator with no result. The overloaded assignment operator makes it possible to use a mathematical notation for algorithms or programs. Thus, the assignment may be defined for types that are not assignment compatible.

The declaration has the form

> **operator** := (FormalOperand1, FormalOperand2);
> Declaration; ...
> **begin**
>   Statement; ...
> **end**;

which is very similar to the declaration of a procedure. The main difference between the above declaration and the declaration of operators described in section 2.7.6 is that there is no result identifier and no type specification. Moreover, the formal operand 1 _must_ be specified by

> **var** Identifier : TypeSecification

whereas the formal operand 2 can be specified by

> **var** { may be omitted }
>   Identifier : TypeSpecification

The algorithm for passing the right side (operand 2) to the left side (operand 1) is usually expressed in the statement part of the assignment operator. In general, the **var**-parameter operand 1 is the parameter returned from this operator.

An overloaded assignment operator is used in the usual assignment statement:

> Variable := Expression

Now, the left and right sides of the assignment statement are to be considered to be assignment compatible according to the type combinations of the overloading (see section 2.3.5).

This new assignment compatibility is not extended to the call by value of subroutines (see section 2.7.10 on page 101).

In the following example dealing with intervals and vectors, we demonstrate how the work with numbers of embedded spaces or the initialization of vectors or matrices is simplified by using overloaded assignment operators.

Example 2.7.17:

```
...
var
    x  : interval;
    iv : ivector[1..n];
    im : imatrix[1..n,1..n];
...
operator := (var x: interval; r: real);    { Op1 }
    begin
        x.inf := r;
        x.sup := r;
    end;
operator := (var iv: ivector; r: real);    { Op2 }
    var i: integer;
    begin
        for i:= lb (iv) to ub (iv) do
            iv[i] := r; { call of Op1 }
    end;
operator := (var im: imatrix; r: real);    { Op3 }
    var i : integer;
    begin
        for i:= lb (im) to ub (im) do
            im[i]:= r; { call of Op2 }
    end;
...
x  :=   5.3;   { call of Op1 delivers point interval }
iv :=   0;     { call of Op2 delivers interval zero vector }
im :=   0;     { call of Op3 delivers interval zero matrix }
```

2.8 Modules

─────────────────────────────────── *PASCAL–XSC* ───────

In Standard PASCAL, a program can only be given as one single program text
that must be completely written before it can be compiled and executed. In
contrast to this, PASCAL–XSC allows the splitting of a program in several parts
called modules which can be developed and compiled separately.

Modules are collections of procedures, functions, operators, affiliated constant
and type definitions, and variable declarations. Modules are declared similarly to
programs, but they are compiled separately. A module has the following syntax:

> **module** name;
>   UseClause; ...
>   **global** declaration; ... { **global** may be omitted }
> **begin** { may be omitted together with the statement part }
>   statement; ...
> **end.**

The module identifier follows the reserved word **module**.

Declarations have the same form as defined in the declaration part of a program. If a declaration is introduced by the reserved word **global**, then all objects
declared in this module are global quantities of this module, i.e. they are available
to be exported into other modules or into the main program. All other declared
objects are local quantities of the module.

In the definition of a global type, the reserved word **global** may occur on the
right-hand side of the equal sign. In this case, the structure of the global type
can also be exported. The type definition

> **global type** complex = **record** re, im: real **end;**

exports the type *complex*, but does not export its structure. Thus, access to the
components of the data structure is only possible within the module containing
the type definition itself. The declaration

> **global type** complex = **global record** re, im: real **end;**

exports both the type identifier *complex* and also the record component identifiers *re, im*.

─────────────────────────────── *PASCAL–XSC* ───

Example 2.8.1:

A simple module definition to provide a complex arithmetic may have the following form:

module ComplexArithmetic;
   global type complex = **global record** re, im: real **end**;
   global operator + (z, w: complex) res: complex;
     begin
       res.re := z.re + w.re;
       res.im := z.im + w.im;
     end;
   global operator * (z, w: complex) res: complex;
     begin
       res.re := z.re * w.re − z.im * w.im;
       res.im := z.re * w.im + z.im * w.re;
     end;
  end.

A program or another module employs a **use**-clause to make visible objects exported from the used modules (import of objects). If the reserved word **global** occurs in a **use**-clause, then all objects being imported by this clause are available for *export* as well. A **use**-clause is defined by:

| **use global**   ModuleIdentifierList   { **global** may be omitted }

Example 2.8.2:

The following module provides an addition for complex vectors on the basis of the module *ComplexArithmetic*:

module ComplexVectorArithmetic;
   use global ComplexArithmetic;
   global type
     complexvector = **global dynamic array** [*] **of** complex;
   global operator + (x, y: complexvector)
             res: complexvector [lbound(x)..ubound(x)];
     var i: integer;
     begin
       for i:= lbound(x) **to** ubound(x) **do**
         res[i] := x[i] + y[i];
     end;
  end.

—————————————————————————————————————— *PASCAL–XSC* ——

If another program or module includes the **use**-clause

use   ComplexVectorArithmetic;

then the type *complexvector* and the appropriate operator + are visible. The
type *complex* and the appropriate operators + and * are also available, since the
defining module is globally linked via

use global   ComplexArithmetic;

That is, the program or module does not need to include the module *ComplexArithmetic* in order to have objects from the *ComplexArithmetic* module
visible. However, if the module *ComplexVectorArithmetic* included only the
clause

use   ComplexArithmetic;

then this clause would also be necessary in the module or in the main program
that imports *ComplexVectorArithmetic*.

   The **use**-clauses build up a module hierarchy among the individual modules
and the main program. The modules may be represented in an acyclic graph that
is similar to a tree structure whose root is represented by the main program. The
modules imported into the main program are given as the children.

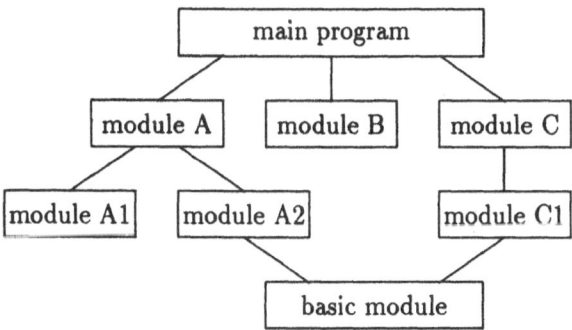

This is not strictly a tree structure since one module (basic module in this figure)
can be used by more than one other module. The **use**-clauses impose a strict
partial order. A mutual or cyclic importation is not allowed, not even indirectly.
The module hierarchy determines the order of the compilation of the modules.
A module may not be compiled until all imported modules have been compiled.
In every module which has to be compiled, at least the exported quantities have
to be declared or defined, although their implementation does not need to be
complete. In the case of procedures for example, empty statement parts are
sufficient.

—————————————————————————————————————— *PASCAL–XSC* ——

Thus, after the first planning stage determining the structure of the individual modules and checking their interfaces, the complete implementation of the modules may be executed in parallel by possibly different teams.

If no more changes are made in the definition part of the modules, i.e. at the declarations of objects being available to be exported, the compilation of the modules is sufficient. Otherwise, all modules which depend on the altered module must be recompiled along with the main program.

A module may contain a statement part after the declaration part. The statement part is executed just once at the beginning of program execution. The statement parts of several modules are executed in an order consistent with the partial ordering of the module hierarchy. In the statement part of a module, the local and global variables can be initialized by using an arbitrary set of statements using the quantities of the module. In the module hierarchy shown in the preceding figure, the statement part of the basic module has to be executed before the statement part of the module A2 which has to occur before the statement part of module A.

2.9   String Handling and Text Processing

Standard PASCAL offers only poor features for processing files of type *text*, characters (*char*), and strings (**packed array** *[1..n]* **of** *char*). Except of the lexical comparisons of strings or characters, there are no expressions involving these types. Manipulation of strings or an input statement for a string variable is not supported.

PASCAL–XSC —————

The dynamic string type (section 2.3.2), the string expression (section 2.4.3.2), and the string functions, comparisons, assignment statements, and input/output of strings support convenient text processing.

The following functions and procedures are provided for the operations which are normally used for text processing:

function image (i: integer) : string;

> Converts the numerical value *i* into a string. Similar to the result from *write (i)* with a current length like the default output format for *integer* values.

function image (i, len: integer) : string;

> Converts the numerical value *i* into a string with a current length *len* (possibly filled by leading blanks). Similar to the output of *integer* values.

function image (r: real) : string;

> Converts the numerical value *r* into a string with a current length like the default output format for *real* values.

function image (r: real; width: integer) : string;

> Converts the numerical value *r* into a string with a current length *width* (possibly filled by leading blanks). Similar to the output of *real* values.

function image (r: real; width, fracs: integer) : string;

> Converts the numerical value *r* into a string with current length *width* (possibly filled with leading blanks) and with *fracs* places after the decimal point. Similar to the output of *real* values.

─────────────────────────────────── *PASCAL–XSC* ───

function image (r: real; width, fracs, round: integer) : string;

Converts the numerical value *r* into a string with a current length *width* and *fracs* places after the decimal point. The value is rounded according to *round*:

$$round \begin{cases} < 0 & \text{rounded downwardly directed} \\ = 0 & \text{rounded to the nearest} \\ > 0 & \text{rounded upwardly directed} \end{cases}$$

function substring (s: string; p, l: integer) : string;

Returns a substring containing *l* characters from *s* starting at position *p*.

function length (s: string) : integer;

Returns the current length of *s*.

function maxlength (**var** s: string) : integer;

Returns the maximum length of the variable *s*.

function pos (sub, s: string) : integer;

Scans the string *s* to find the first occurrence of *sub* in *s*. If the pattern is not found, *pos* returns the value 0.

function ival (s: string) : integer;

Converts the first part of the string *s*, which represents a numeric value according to the rules of *integer* constants, into an *integer* value. Leading blanks as well as trailing characters are neglected.

function ival (s: string; **var** rest: string) : integer;

Converts the first part of the string *s*, which represents a numeric value according to the rules of *integer* constants, into an *integer* value. Leading blanks are neglected, whereas trailing characters are passed back in the string *rest*.

function rval (s: string) : real;

Converts the first part of the string *s*, which represents a numeric value according to the rules of *real* constants, into a *real* value. Leading blanks as well as trailing characters are neglected.

PASCAL–XSC

function rval (s: string; **var** rest: string) : real;

> Converts the first part of the string *s*, which represents a numeric value according to the rules of *real* constants, into a *real* value. Leading blanks are neglected, whereas trailing characters are passed back in the string *rest*.

function rval (s: string; round: integer) : real;

> Converts the first part of the string *s*, which represents a numeric value according to the rules of *real* constants, into a *real* value rounded according to *round* (see *image*). Leading blanks as well as trailing characters are neglected.

function rval (s: string; round: integer; **var** rest: string) : real;

> Converts the first part of the string *s*, which represents a numeric value according to the rules of *real* constants, in a *real* value rounded according to *round* (see *image*). Leading blanks are neglected, whereas trailing characters are passed back in the string *rest*.

procedure setlength (**var** s: string; len: integer);

> Sets the current length of the string variable *s* to *len*. The value *len* must lie within the range 0..*maxlength(s)*.

Example 2.9.1:

image (4728,5)	delivers	' 4728'
image (3.14159,7,4,1)	delivers	' 3.1416'
substring ('AAABB',3,3)	delivers	'ABB'
length ('abcde')	delivers	5
pos ('AB','AAABB')	delivers	3
ival ('512')	delivers	512
rval ('−1.5E6')	delivers	−1.5E+06

The relational operators

$$=, \quad <>, \quad <=, \quad <, \quad >=, \quad >$$

applied to strings have their usual meaning derived from lexical ordering. For a string *s1* with the length n and a string *s2* with a length $m > n$, both coinciding on the first m positions, the comparison *s1* < *s2* is true.

The additional operator

> **in**

for two string operands tests substrings. The expression *s1* **in** *s2* delivers *true* if *s1* is a substring of *s2* and *false* otherwise.

Example 2.9.2:

> Let   s5 := 'AAABB';
> then   'A' in s5   delivers *true*
> and   'BBA' in s5   delivers *false*.

The use of assignment statement

| StringVariable := StringExpression

is always possible if the string variable is of type *string* and if the string expression is of type *string* or of any array string type.

> If the actual length of the string expression exceeds the maximum length of the variable, the extra characters on the end of the string expression are truncated.

Example 2.9.3:

> The following types and variables are given:

> **type**   string_10 = string[10];
>   string_20 = string[20];
> **var**   s5   :   string[5];
>   s10   :   string_10;
>   s20   :   string_20;
>   s   :   string; { length implementation-dependent }

> Then it is possible to write:

> s5   :=   'ABCDE';
> s10   :=   s5;
> s20   :=   'AABBCC';
> s5   :=   s20; { s5 contains the value 'AABBC' }
> s20   :=   ''; { empty string }
> s5   :=   'AAA' + 'BBB' { s5 contains the value 'AAABB' }

2.9.1   Input of Characters and Strings

While the output of characters and strings in PASCAL is processed according to
the programmer's intention, the entering of these types from the console very often
produces unexpected results.

To read in a *char* variable c using the statement

    read (c);

the statements

    c := input↑;
    get (input);

are executed according to the definition of *read* (see section 2.5.2). With the first
read on *input*, a blank is assigned to the variable c. This blank corresponds to the
end-of-line character, since immediately after the start of the program, *eoln (input)*
= *true* .

Example 2.9.4:

The program

        program testread1 (input, output);
        var c : char;
        begin
          read (c);
          writeln (c);
        end.

would input the end-of-line character and output a blank without accepting
any input via the user's console.

Hence, we have to take special care of the end-of-line character while reading characters, in contrast to the input of *integer* or *real* numbers which neglects the end-of-line
character as it neglects a blank. Appropriate use of *readln* or *get* (see section 2.5.2)
is necessary.

If we use a procedure *read_char* declared by

        procedure read_char (**var** f: text; **var** c: char);
        begin
          if eoln (f) **then**
            readln (f);
          read (f, c);
        end;

we can read a character (not equal to the end-of-line character) without having to
worry about end-of-line arrangements or unexpected effects.

Example 2.9.5:

The following program enables the user to enter a character, which is printed immediately afterwards.

```
program testread2 (input, output);
var c : char;
procedure read_char ...
{ same procedure declaration as above }
...
begin
   read_char (input, c);
   writeln (c);
end.
```

─────────────────────────────────── *PASCAL–XSC* ───────

The particularity of PASCAL concerning the input of characters also applies to the input of strings in PASCAL–XSC. In the following tables, some examples illustrate this fact. In these tables, ↵ denotes the *Return* key for input from the console or the end-of-line character for input from a file, and ⊔ denotes the blank character. The variables *S5* and *S10* are defined as strings with a maximum length 5 and 10, respectively. The file variable *f* is of type *text*.

String Input from Console

Statements	Input	Output
read (S5, S10) writeln (S5); writeln (S10);	*not possible*	 ⊔ ⊔
readln (S5, S10) writeln (S5); writeln (S10);	ABCDEFGHIJKLMNO↵	 ⊔ ⊔
readln; read (S5, S10); writeln (S5); writeln (S10);	ABCDEFGHIJKLMNO↵	 ABCDE FGHIJKLMNO
readln; readln (S5, S10); writeln (S5); writeln (S10);	ABCDE↵ FGHIJKLMNO↵	 ABCDE ⊔
readln; read (S5); readln; read (S10); writeln (S5); writeln (S10);	ABCDE↵ FGHIJKLMNO↵	 ABCDE FGHIJKLMNO

PASCAL–XSC

String Input from File

Statements	File Contents	Output
read (f, S5, S10); writeln (S5); writeln (S10);	ABCDEFGHIJKLMNO	ABCDE FGHIJKLMNO
read (f, S5, S10); writeln (S5); writeln (S10);	ABCDE FGHIJKLMNO	ABCDE ⊔
readln (f, S5); readln (f, S10); writeln (S5); writeln (S10);	ABCDE FGHIJKLMNO	ABCDE FGHIJKLMNO

For the type *string*, the appropriate use of *readln* avoids the unexpected input of the end-of-line character. For instance, we can use an overloaded procedure *read* (see section 2.7.11) declared by:

```
procedure read (var f : text; var s: string);
var c : char;
begin
  if eoln (f) then
    readln (f);
  s := ''; { empty string }
  while not eoln (f) do
  begin
    read (f, c);
    s := s + c;
  end;
end;
```

For an arbitrary text file *f* and a *string* variable *s*, this procedure can be applied in the forms

read (s);   read (input, s);   read (f, s);

to read in dynamic strings line by line (due to overloading of *read* and *write* as described in section 2.7.11).

Example 2.9.6:

Let us write a PASCAL–XSC program to convert German text written in lower-case letters containing the strings (umlauts) **ae**, **oe**, and **ue**, into a form which can serve as input for the text system LATEX using the document style option *german*. It is necessary to replace:

ae by "a

oe by "o

ue by "u

For simplicity, special cases like **aee** may be neglected. Furthermore, the word PASCAL should be changed into Pascal and marked for typing in boldface by enclosing the word in the form:

{\bf Pascal}

The following PASCAL–XSC program enters the text from the text file *texin.txt*, processes the changes, writes the changed text to the text file *texout.txt*, and terminates.

```
program umlauts (output, infile, outfile);
operator ** (line, umlaut: string) res : string;
    { Replaces the umlauts contained in the   }
    { line by the corresponding TeX sequence. }
    var
        p : integer;
    begin
        p := pos (umlaut, line);
        while (p > 0) do
        begin
            line[p] := '"';
            line[p+1] := umlaut[1];
            p := pos (umlaut, line);
        end;
        res := line;
    end;
var
    line, help1, help2 : string;
    infile, outfile : text;
    len, position : integer;
```

────────────────────────── *PASCAL–XSC* ──────────

```
begin
   reset (infile, 'texin.txt');
   rewrite (outfile, 'texout.txt');
   while not eof (infile) do
   begin
      readln (infile, line);
      line := line ** 'ae';
      line := line ** 'oe';
      line := line ** 'ue';
      len := length (line);
      position := pos ('PASCAL', line);
      while (position > 0) do
      begin
         help1:= substring (line, 1, position-1) + '{\bf Pascal}';
         help2:= substring (line, position+6, len−position−5);
         line:= help1 + help2;
         len:= length (line);
         position:= pos ('PASCAL', line);
      end;
      writeln (outfile, line);
   end;
end.
```

2.10   How to Use Dynamic Arrays

─────────────────────────────── *PASCAL–XSC* ───────

True dynamic allocation of array lengths can only occur when declaring dynamic array variables within procedures or functions as described in section 2.3.2. In the declaration part of these routines, global quantities or formal parameters are used in the expressions for the index bounds of the arrays. In the body of the main program, only constants, imported variables, or expressions that can be evaluated at the point of the declaration may be used in the index expressions.

An experienced programmer might be able to realize full dynamic array lengths in the main program by using a special module initialization part or function calls for the index bounds. Nevertheless, in this section we discuss the usual manner of working with dynamic arrays. Usually, the original main program, which works with dynamic arrays, is moved into a procedure or function. The body of the new main program then consists only of the entering of values which are necessary for the calculation of the index bounds and of the call of the new "main procedure" or "main function".

The template of a PASCAL–XSC program which uses dynamic arrays is

```
program dynprog (input, output);
type
   dyntype = dynamic array [*] of comptype;
   { further declarations }
   ...
var
   low, upp: integer;
   { further declarations }
   ...
procedure main (low, upp: integer);
   var
      a, b, c: dyntype [low..upp];
      { further declarations}
      ...
   begin
      { main program, moved into the procedure }
      ...
   end;

begin { new main program}
   read (low,upp);
   main (low,upp);
end.
```

───────────────────────────── PASCAL–XSC ───

In the new main program, the procedure *main* could also be called within a loop, within which new index bounds *low* and *upp* are entered. This might be useful in an algorithm which improves a computed result by enlarging the dimension of the dynamic arrays employed.

Example 2.10.1:

```
program longnumber (input, output);
type
    long = dynamic array [*] of real;
var
    len: integer;
    ...
function ok (len: integer) : boolean;
    var
        lz1, lz2, lz3: long [1..len];
        erg: real;
        ...
    begin
        { algorithm }
        ...
        writeln ('result using length ', len:1, ': ', erg);
        if { precision of res ok } then
            ok := true
        else
            ok := false;
    end;

begin
    repeat
        write ('length of type long: ');
        read (len);
    until ok (len);
end.
```

As a final example for the handling of dynamic arrays, we list a program to compute the transposed matrix for arbitrary (square or rectangular) matrices of arbitrary dimension.

Example 2.10.2:

```
program transpose (input,output);
type
    matrix = dynamic array [*,*] of real;
function transp (var a: matrix) :
        matrix [lbound(a,2)..ubound(a,2), lbound(a,1)..ubound(a,1)];
    var
        i, j: integer;
    begin
        for i:=lbound (a,1) to ubound (a,1) do
            for j:=lbound (a,2) to ubound (a,2) do
                transp[j,i] := a[i,j];
    end;
procedure main (no_of_rows, no_of_columns: integer);
    var
        i, j: integer;
        A: matrix [1..no_of_rows,1..no_of_columns];
        T: matrix [1..no_of_columns,1..no_of_rows];
    begin
        writeln ('Enter the matrix elements of A (row by row) ');
        for i:=1 to no_of_rows do
            for j:=1 to no_of_columns do
                read(A[i,j]);
        writeln ('Transposed matrix of A:');
        T:= transp(A);
        for i:=1 to no_of_columns do
        begin
            for j:=1 to no_of_rows do
                write (T[i,j]);
            writeln;
        end;
    end;
var
    no_of_rows, no_of_columns: integer;
begin
    writeln ('Size of A:');
    write ('Number of rows: ');
    read (no_of_rows);
    write ('Number of columns: ');
    read (no_of_columns);
```

```
  while (no_of_rows > 0) and (no_of_columns > 0) do
  begin
    main (no_of_rows,no_of_columns);
    writeln ('Size of A:');
    write ('Number of rows: ');
    read (no_of_rows);
    write ('Number of columns: ');
    read (no_of_columns);
  end;
end.
```

Chapter 3

The Arithmetic Modules

Numerical methods require computations not only in the space of real numbers, but also with complex numbers, and vectors and matrices over these numbers (see [1], [2], [19], or [33]). To fulfill all these requirements, PASCAl–XSC provides the corresponding types with the necessary operators and functions.

All arithmetic operators are of *maximum accuracy* as described in section 1.3 or for *real* operations in section 2.4.1.2. The result is computed to at least 1 ulp accuracy.

PASCAL–XSC provides a complete expression concept for the additional numerical types

complex	for	complex numbers
interval	for	real intervals
cinterval	for	complex intervals
rvector	for	real vectors
cvector	for	complex vectors
ivector	for	interval vectors
civector	for	complex interval vectors
rmatrix	for	real matrices
cmatrix	for	complex matrices
imatrix	for	interval matrices
cimatrix	for	complex interval matrices

This expression concept is not restricted to operands of the same type. Moreover, almost every operation which is usually applied to different operand types in the mathematics is provided. Therefore, more than 1000 arithmetic operators are provided. In addition, PASCAL–XSC enables the user to form logical expressions with these types by providing a comparably large number of relational operators. This large number of operators and functions makes it possible to transfer mathematical computations of engineering and science into a clearly structured programming code.

In most cases, the original theoretical formulas or algorithms can be used as program parts with only few changes. This fact is supported by predefined overloadings of the assignment operator :=.

The following table 1 is a survey of the predefined arithmetic operators for the arithmetic types.

Table 1: Predefined Arithmetical Operators

right operand / left operand	integer real complex	interval cinterval	rvector cvector	ivector civector	rmatrix cmatrix	imatrix cimatrix
[1])	+,−	+,−	+,−	+,−	+,−	+,−
integer real complex	o,o<,o>,[2)] +*	+,−,*,/, +*	*,*<,*>	*	*,*<,*>	*
interval cinterval	+,−,*,/, +*	+,−,*,/, +*,**	*	*	*	*
rvector cvector	*,*<,*>, /,/<,/>	*,/	o,o<,o>,[3)] +*	+,−,*,[4)] +*		
ivector civector	*,/	*,/	+,−,*,[4)] +*	+,−,*,[4)] +*,**		
rmatrix cmatrix	*,*<,*>, /,/<,/>	*,/	*,*<,*>	*	o,o<,o>,[3)] +*	+,−,*,[4)] +*
imatrix cimatrix	*,/	*,/	*	*	+,−,*,[4)] +*	+,−,*,[4)] +*,**

[1]) The operators of this row are monadic (i.e. there is no left operand).

[2]) $o \in \{+,-,*,/\}$

[3]) $o \in \{+,-,*\}$, where * denotes the scalar or matrix product.

[4]) The * denotes the scalar or matrix product.

+* : Interval hull (smallest interval enclosing both operands)

** : Interval Intersection

Remark: The block of table 1 which is marked by [2]) contains the operators for *real* and *integer* operands of Standard PASCAL. The operators * (scalar product), +* (interval hull), and ** (interval intersection) are provided in the corresponding matrix/vector modules and interval modules, respectively.

Table 2 gives an overview of the relational operators for the arithmetic types available in PASCAL–XSC.

Table 2: Predefined Relational Operators

left operand \ right operand	integer real complex	interval cinterval	rvector cvector	ivector civector	rmatrix cmatrix	imatrix cimatrix
integer real complex	=,<>, <=,<, >=,>	in =,<>				
interval cinterval	=,<>	in,><,[1]) =,<>, <=,<, >=,>				
rvector cvector			=,<>, <=,<, >=,>	in =,<>		
ivector civector			=,<>	in,><,[1]) =,<>, <=,<, >=,>		
rmatrix cmatrix					=,<>, <=,<, >=,>	in =,<>
imatrix cimatrix					=,<>	in,><,[1]) =,<>, <=,<, >=,>

[1]) The operators <= and < denote the "subset" relation; >= and > denote the "superset" relation

>< : Test for disjointedness of intervals

in :   Test for membership of a point in an interval or test for strict inclusion of an interval in the interior of an interval

The large number of operators are provided in special arithmetic modules which contain the operators listed above and a set of predefined functions. For the types *complex, interval,* and *cinterval,* this set contains all mathematical functions which are provided for type *real* (see section 2.4.1.2).

Hence, the following modules are available:

C_ARI	complex arithmetic
I_ARI	interval arithmetic
CI_ARI	complex interval arithmetic
MV_ARI	real matrix/vector arithmetic
MVC_ARI	complex matrix/vector arithmetic
MVI_ARI	interval matrix/vector arithmetic
MVCI_ARI	complex interval matrix/vector arithmetic

Each of these modules is described in the following sections. All types, operators, transfer functions, overloadings of :=, predefined arithmetic functions, and input/output procedures are systematically explained. The domains and ranges of the functions are implementation-dependent. The rules of overloading for *read* and *write* described in section 2.7.11 apply to the input/output procedures of the modules, i.e. they can be used with an optional file parameter and with an arbitrary number of input/output parameters. Therefore, the description of these procedures is restricted to the explanation of the possible input and output formats.

These shortened module names are chosen due to the implementation-dependent maximum number of significant characters in the module name which must be equal to the corresponding file name. Some systems have special requirements concerning the length of file names or the length of entry names for linker interfaces. With these short names, modules are portable across all systems.

Definition of the Arithmetic Operators

The type of the result of scalar arithmetic operations is defined in the mathematical sense according to the following hierarchy of types:

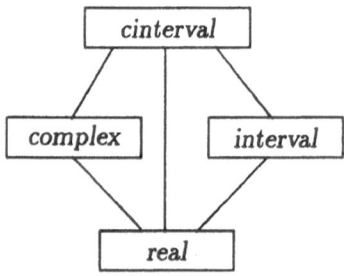

The result is always the lowest type containing both operand types.

For the matrix/vector operations, the structure of the result follows from the structures of the operands:

$$
\begin{array}{ccccc}
v & + & v & = & v \\
m & + & m & = & m \\
\\
v & - & v & = & v \\
m & - & m & = & m \\
\\
v & / & s & = & v \\
m & / & s & = & m \\
\end{array}
\qquad\qquad
\begin{array}{ccccc}
v & * & v & = & s \\
s & * & v & = & v \\
v & * & s & = & v \\
m & * & v & = & v \\
s & * & m & = & m \\
m & * & s & = & m \\
m & * & m & = & m \\
\end{array}
$$

Structure of the Result Type for Matrix/Vector Operations

s = scalar,   v = vector,   m = matrix

The type of the result follows from the above hierarchy of types depending on the two component types.

All matrix/vector operations assume the number of corresponding components of the operands is equal, i.e. the corresponding index ranges must have the same length. The index ranges themselves may be different as in

```
var
   p : complex;
   a : rvector[1..10];
   b : cvector[11..20];
...
p := a * b;
```

In the definitions of the operators in the following sections, the descriptions given assume that the index ranges are identical. For example, the scalar product $p = a * b$ is described as

```
#* (for i:= lb(a) to ub(a) sum
       (a[i] * b[i]) )
```

but it is implemented as

```
#* (for i:= lb(a) to ub(a) sum
       (a[i] * b[i + lb(b) − lb(a)]) )
```

Definition of the Relational Operators

The relational operators in these modules are based upon the set of relational operators for the type *real*. This set of operators is used to define the operators <= and = for a structured numerical data type SNDT (Structured Numerical Data Type). The operator = is implemented in such a manner that it delivers *true* if and only if all components of the SNDT fulfill the equality. The definition of the operator <=

depending on the type of the operands is explained in the corresponding section of the defining modules.

All further relational operators for elements $a, b \in$ SNDT are defined by:

$$(RD) \quad \begin{array}{llll} a & <> & b & := & \text{not } (a = b) \\ a & < & b & := & (a <= b) \text{ and } (a <> b) \\ a & > & b & := & b < a \\ a & >= & b & := & b <= a \end{array}$$

Overloading of the Assignment Operator

The assignment operator is overloaded to provide several basic type conversions and some array type initializations. All type conversions are defined according to the mathematical embedding (for example, the real numbers are embedded in the complex numbers), i.e. the value is invariant. Initializations by means of assignments of scalar types to vector or matrix types are defined componentwise, i.e. the same value is assigned to all components of the array type. For all these overloaded assignment operators, there are *no* rounding or conversion errors.

Nevertheless, we have to remember the special problematic nature of conversion described in section 2.3.1 in connection with literal constants on the right side of the assignment. For the reasons described in section 2.3.1, a real constant is converted into the internal *real* format *before* the assignment is executed.

Accuracy of the Predefined Functions

All complex functions deliver results of at least 1 ulp accuracy.

The interval functions always compute a floating-point interval that contains the exact interval result. In most cases, the smallest enclosing interval is computed, but there are some special cases in which the bounds differ by 2 ulp.

The complex interval functions achieve the same accuracies for their real and imaginary parts.

3.1   The Module C_ARI

Complex Arithmetic

This module supplies all operators, functions, and procedures necessary for computations with complex numbers.

Type

The type *complex* defined by

type complex = **record** re, im : real **end**;

is part of the language core of PASCAL–XSC. It is based upon the cartesian representation of a complex number z in the form

$$z = x + iy,$$

where x denotes the real part and y denotes the imaginary part of z.

Operators

All predefined arithmetic operators of this module deliver the result type *complex*. There are the monadic operators $+, -$ and the four basic operations $+, -, *, /$, each with three different kinds of rounding. All roundings are interpreted componentwise.

The relational operators $=, <>, <, <=, >, >=$ are defined on the base of $=$ and $<=$ according to (RD) (see page 130). If a and b are of type *complex* then

$$a <= b \iff (a.re <= b.re) \textbf{ and } (a.im <= b.im).$$

Comparison with an *integer* or *real* operand is allowed as well.

left operand \ right operand	integer real	complex
monadic		$+, -$
integer real		o v
complex	o v	o v

The Operators of Module C_ARI

$$o \in \{+, +<, +>, -, -<, ->, *, *<, *>, /, /<, />\}$$

$$v \in \{=, <>, <, <=, >, >=\}$$

Transfer Functions

The following transfer functions are provided for type conversions between the types *real* and *complex*:

Function	Result Type	Meaning
compl (r1,r2)	*complex*	Complex number with real part *r1* and imaginary part *r2*
compl (r)	*complex*	Complex number with real part *r* and imaginary part 0
re (c)	*real*	Real part of *c*
im (c)	*real*	Imaginary part of *c*

r, r1, r2 = *real* Expression,   c = *complex* Expression

Example 3.1.1:

The imaginary unit *i* can be generated using the expression

compl (0,1) .

Overloading of the Assignment Operator

The type conversion *real* to *complex* is provided as an overloading of the assignment operator:

Assignment	Meaning
c := r	c := compl (r)

c = *complex* variable
r = *real* expression

Predefined Functions

All mathematical functions of PASCAL–XSC available for *real* arguments are also supplied for complex arguments. Moreover, functions for the computations of the angle component φ of the exponential representation $z = r \cdot e^{i\varphi}$ of a complex number z and for the conjugation (reflection about the real axis) are provided.

Function	Result Type		Meaning
sqr (c)	*complex*	$c^2 = c \cdot c$	Square
sqrt (c)	*complex*	$\sqrt{c}$	Square Root (Real part > 0)
exp (c)	*complex*	e^c	Exponential Function
exp2 (c)	*complex*	2^c	Power Function, Base 2
exp10 (c)	*complex*	10^c	Power Function, Base 10
ln (c)	*complex*	$\ln (c)$	Natural Logarithm
log2 (c)	*complex*	$\log_2(c)$	Logarithm, Base 2
log10 (c)	*complex*	$\log_{10}(c)$	Logarithm, Base 10
sin (c)	*complex*	$\sin (c)$	Sine
cos (c)	*complex*	$\cos (c)$	Cosine
tan (c)	*complex*	$\tan (c)$	Tangent
cot (c)	*complex*	$\cot (c)$	Cotangent
arcsin (c)	*complex*	$\arcsin (c)$	Arc Sine
arccos (c)	*complex*	$\arccos (c)$	Arc Cosine
arctan (c)	*complex*	$\arctan (c)$	Arc Tangent
arccot (c)	*complex*	$\text{arccot} (c)$	Arc Cotangent
sinh (c)	*complex*	$\sinh (c)$	Hyperbolic Sine
cosh (c)	*complex*	$\cosh (c)$	Hyperbolic Cosine
tanh (c)	*complex*	$\tanh (c)$	Hyperbolic Tangent
coth (c)	*complex*	$\coth (c)$	Hyperbolic Cotangent
arsinh (c)	*complex*	$\text{arsinh} (c)$	Inverse Hyperbolic Sine
arcosh (c)	*complex*	$\text{arcosh} (c)$	Inverse Hyperbolic Cosine
artanh (c)	*complex*	$\text{artanh} (c)$	Inverse Hyperbolic Tangent
arcoth (c)	*complex*	$\text{arcoth} (c)$	Inverse Hyperbolic Cotangent
conj (c)	*complex*	$\bar{c} = x - iy$	Conjugation of $c = x + iy$
arg (c)	*real*	φ	Argument of $c = r \cdot e^{i\varphi}$
abs (c)	*real*	$r = \sqrt{x^2 + y^2}$	Absolute value of $c = r \cdot e^{i\varphi} = x + iy$

c = *complex* expression

The domains and ranges of the predefined functions are implementation-dependent and are described in the user manual.

Input/Output Procedures

This module supplies the procedures

procedure read (**var** f: text; **var** a: complex);
procedure write (**var** f: text; a: complex);

with optional file parameters, arbitrarily many input/output parameters, but without format specifications.
A complex number $c = x + iy$ must be entered in the form

(x, y)

or in the form

x.

In the second case, the imaginary part y is set to 0. x and y are real constants that are rounded to the nearest floating-point numbers. The output of a complex number rounds both the real and imaginary parts to the nearest decimal numbers. It is displayed in the form

(x, y)

with an implementation-dependent default format for the *real* values x and y.

Example 3.1.2:

If c is of type *complex*, then the statements

read (c);
writeln (c);

accept the input data

$-1.23456789,$

and write the *complex* data

$(-1.234567890000E+00, 0.000000000000E+00).$

Another *real* representation may be used depending on the implementation.

3.2   The Module I_ARI

Interval Arithmetic

This module supplies all operators, functions, and procedures necessary for computations with intervals.

Type

The type *interval* defined by

   type interval = **record** inf, sup : real **end**;

is part of the language core of PASCAL–XSC. It is based upon the representation of a real interval x in the form

   $x = [x_{inf}, x_{sup}]$

representing the set $\{y \in \mathbb{R} | x_{inf} \leq y \leq x_{sup}\}$. *inf* denotes the infimum (lower bound), and *sup* denotes the supremum (upper bound) of x.

Operators

All predefined arithmetic and lattice operators deliver the result type *interval*. There are the monadic operators $+, -$ and the four basic operations $+, -, *, /$, each with the rounding to the smallest enclosing interval. The relational operators $=, <>, <, <=, >, >=$ are to be interpreted as the corresponding set operators. Their meaning is

=	equal
<>	not equal
<	proper subset
<=	subset
>	proper superset
>=	superset

These operators are defined on the base of = and <= according to (RD) (see page 130). If x and y are of type *interval*, then

   x <= y $\Longleftrightarrow$ (x.inf >= y.inf) **and** (x.sup <= y.sup).

Moreover, this module supplies the operator **in** for the relation "is contained in" or "is contained in the interior" between a *real*- and an *interval* operand or between two *interval* operands. The operator **in** satisfies

   x **in** y $\Longleftrightarrow$ (x.inf > y.inf) **and** (x.sup < y.sup).

The operator $><$ tests for disjointedness of two intervals. Two intervals x and y are disjoint if $x \cap y = \emptyset$ (empty set). The lattice operators $+*$ and $**$ denote the interval hull and the interval intersection, respectively. The operator $+*$ delivers the smallest interval enclosing both operands. The operator $**$ delivers the intersection. It is an error if the intersection is empty.

left operand \ right operand	integer real	interval
monadic		$+, -$
integer real	$+*$	$\diamond$ $in, =, <>$ $+*$
interval	$\diamond$ $=, <>$ $+*$	$\diamond$ $in, \vee, ><$ $+*, **$

The Operators of Module I_ARI

$$\diamond \in \{+, -, *, /\}$$

$$\vee \in \{=, <>, <, <=, >, >=\}$$

Example 3.2.1:

If a and b of type *interval* are defined as

$$a = [-1,3]$$
$$b = [3,4],$$

then the operators $+, -, *, ><, +*,$ and $**$ yield the results:

Expression			Result
a	$+$	b	$[2,7]$
a	$-$	b	$[-5,0]$
a	$*$	b	$[-4,12]$
a	$+*$	b	$[-1,4]$
a	$**$	b	$[3,3]$
a	$><$	b	*false*

Transfer Functions

The following transfer functions are provided for type conversions between the types *real* and *interval*:

Function	Result Type	Meaning
intval (r1,r2)	*interval*	Interval with *inf* = r1 and *sup* = r2 [*]
intval (r)	*interval*	Interval with *inf* = *sup* = r
inf (i)	*real*	Lower bound of *i*
sup (i)	*real*	Upper bound of *i*

r, r1, r2 = *real* expression,   i = *interval* expression

[*]: r1 <= r2 is assumed, otherwise an error occurs.

Overloading of the Assignment Operator

The type conversion *real* to *interval* is provided as an overloaded assignment operator:

Assignment	Meaning
i := r	i := intval (r)

i = *interval* variable

r = *real* expression

Predefined Functions

All mathematical functions of PASCAL–XSC available for real arguments are supplied for interval arguments *i*. These interval functions F satisfy $F(i) \supseteq f(i) = \{f(r) : r \in i\}$. Moreover, functions for the computation of the midpoint and diameter of intervals are available. In connection with enclosure methods, the function *blow* is provided for the *epsilon inflation* (see [46]).

Function	Result Type		Meaning
sqr (i)	*interval*	$i^2 = \{r^2 : r \in i\}$	Interval Square
sqrt (i)	*interval*	$\sqrt{i}$	Square Root
exp (i)	*interval*	e^i	Exponential Function
exp2 (i)	*interval*	2^i	Power Function, Base 2
exp10 (i)	*interval*	10^i	Power Function, Base 10
ln (i)	*interval*	$\ln (i)$	Natural Logarithm
log2 (i)	*interval*	$\log_2(i)$	Logarithm, Base 2
log10 (i)	*interval*	$\log_{10}(i)$	Logarithm, Base 10

i = *interval* expression

Function	Result Type	Meaning	
sin (i)	*interval*	sin (*i*)	Sine
cos (i)	*interval*	cos (*i*)	Cosine
tan (i)	*interval*	tan (*i*)	Tangent
cot (i)	*interval*	cot (*i*)	Cotangent
arcsin (i)	*interval*	arcsin (*i*)	Arc Sine
arccos (i)	*interval*	arccos (*i*)	Arc Cosine
arctan (i)	*interval*	arctan (*i*)	Arc Tangent
arctan2 (i1,i2)	*interval*	arctan (*i1/i2*)	Arc Tangent
arccot (i)	*interval*	arccot (*i*)	Arc Cotangent
sinh (i)	*interval*	sinh (*i*)	Hyperbolic Sine
cosh (i)	*interval*	cosh (*i*)	Hyperbolic Cosine
tanh (i)	*interval*	tanh (*i*)	Hyperbolic Tangent
coth (i)	*interval*	coth (*i*)	Hyperbolic Cotangent
arsinh (i)	*interval*	arsinh (*i*)	Inverse Hyperbolic Sine
arcosh (i)	*interval*	arcosh (*i*)	Inverse Hyperbolic Cosine
artanh (i)	*interval*	artanh (*i*)	Inverse Hyp. Tangent
arcoth (i)	*interval*	arcoth (*i*)	Inverse Hyp. Cotangent
abs (i)	*interval*	$\lvert i \rvert = \{\lvert r \rvert : r \in i\}$	Absolute Value
mid (i)	*real*	$m = \#*(0.5*\inf(i) + 0.5*\sup(i))$	Midpoint of *i*
diam (i)	*real*	$d = \sup(i) -> \inf(i)$	Diameter of *i*
blow (i,r)	*interval*	‡	Epsilon Inflation

$$i, i1, i2 = \textit{interval expression} \quad r, m, d = \textit{real expression}$$

‡ :   y   :=   (1 + r) * i − r * i;
      blow   :=   intval (pred(inf(y)), succ(sup(y)));

The domains and ranges of the predefined functions are implementation-dependent and are described in the user manual.

Example 3.2.2:

If a and b of type *interval* are defined as

    a := intval (−1,3)
    b := intval (2)

then the functions *abs*, *sqr*, *mid*, and *diam* deliver the results:

Expression	Result
abs (a)	[0,3]
abs (b)	[2,2]
sqr (a)	[0,9]
sqr (b)	[4,4]
mid (a)	1
diam (a)	4

Input/Output Procedures

This module supplies the procedures

>**procedure** read (**var** f: text; **var** a: interval);
>**procedure** write (**var** f: text; a: interval);

with optional file parameters, arbitrarily many input/output parameters, but without format specifications.

An interval $i = [x, y]$ must be entered in the form

>$[x, y]$

or in the form

>x.

In the first case, the values of x and y are rounded to the next-smaller and the next-larger floating-point number, respectively (i.e. rounding to the smallest enclosing interval). The second case is a simplified notation for $i = [x, x]$. If x is not exactly representable, the smallest interval enclosing x is generated.

The output of an interval is done with interval rounding (x rounded downwardly, y rounded upwardly) in the form

>$[x, y]$

with an implementation-dependent default format for the *real* values x and y.

Example 3.2.3:

>If *int* is of type *interval*, then the statements

>>read (int);
>>writeln (int);

>accept the input data

>>0.245,

>and write the interval

>>[2.4499...99E−001, 2.4500...01E−001],

>if 0.245 is not exactly representable. Another *real* representation may be used depending on the the implementation.

3.3   The Module CI_ARI

Complex Interval Arithmetic

This module supplies all operators, functions, and procedures necessary for computations with complex intervals.

Type

The type *cinterval* defined by

type cinterval = **record** re, im : interval **end**;

is part of the language core of PASCAL–XSC. It is based upon the usual representation of a complex interval z in the form

$$z = [x_{inf}, x_{sup}] + i \cdot [y_{inf}, y_{sup}]$$

representing a rectangle in the complex plane (rectangular interval).

Operators

All predefined arithmetic and lattice operators deliver the result type *cinterval*.
There are the monadic operators $+, -$ and the four basic operations $+, -, *, /$, each
with the rounding to the smallest enclosing complex interval. The relational operators $=, <>, <, <=, >, >=$ are to be interpreted as the corresponding set operators.
Their meaning is

=	equal
<>	not equal
<	propper subset
<=	subset
>	propper superset
>=	superset

These operators are defined on the base of $=$ and $<=$ according to (RD) (see page
130). If v and w are of type *cinterval*, then

v <= w $\Longleftrightarrow$ (v.re <= w.re) **and** (v.im <= w.im).

The operators on the right side of the equivalence are the ones for intervals.

Moreover, this module supplies the operator in for the relations "is contained
in" and "is contained in the interior" For two complex intervals v and w, the operator
in satisfies

v in w $\Longleftrightarrow$ (v.re in w.re) **and** (v.im in w.im).

The operator $><$ tests for disjointedness of two complex intervals. Two complex intervals v, w are disjoint if $v \cap w = \emptyset$. The lattice operators $+*$ and $**$ denote the complex interval hull and the complex interval intersection, respectively. The operator $+*$ delivers the smallest complex interval enclosing both operands. The operator $**$ delivers the intersection. It is an error if the intersection is empty.

left Operand \ right Operand	integer real	complex	interval	cinterval
monadic				$+, -$
integer real		$+*$		$\diamond$ in, $=$, $<>$ $+*$
complex	$+*$	$+*$	$\diamond$ in, $=$, $<>$ $+*$	$\diamond$ in, $=$, $<>$ $+*$
interval		$\diamond$ $=$, $<>$ $+*$		$\diamond$ in, $\lor$, $><$ $+*$, $**$
cinterval	$\diamond$ $=$, $<>$ $+*$	$\diamond$ $=$, $<>$ $+*$	$\diamond$ $\lor$, $><$ $+*$, $**$	$\diamond$ in, $\lor$, $><$ $+*$, $**$

The Operators of Module CLARI

$$\diamond \in \{+, -, *, /\}$$

$$\lor \in \{=, <>, <, <=, >, >=\}$$

Example 3.3.1:

If *ca* of type *cinterval* is

$$ca = [-1,3] + i \, [3,4],$$

then the operators $+$, $-$, and $*$ deliver

Expression	Result
ca $+$ ca	$[-2,6] + i \, [6,8]$
ca $-$ ca	$[-4,4] + i \, [-1,1]$
ca $*$ ca	$[-19,0] + i \, [-8,24]$

Transfer Functions

The following transfer functions are provided for type conversions between the types *real*, *complex*, *interval*, and *cinterval*:

Function	Result Type	Meaning
compl (i1,i2)	cinterval	Complex interval with real part i1 and imaginary part i2
compl (r,i)	cinterval	Complex interval with real part r and imaginary part i
compl (i,r)	cinterval	Complex interval with real part i and imaginary part r
compl (i)	cinterval	Complex interval with real part i and imaginary part 0
intval (c1,c2)	cinterval	Complex interval with real part [c1.re,c2.re] and imaginary part [c1.im,c2.im] (*)
intval (r,c)	cinterval	Complex interval with real part [r,c.re] and imaginary part [0,c.im] (*')
intval (c,r)	cinterval	Complex interval with real part [c.re,r] and imaginary part [c.im,0] (*")
intval (c)	cinterval	Complex interval with real part [c.re,c.re] and imaginary part [c.im,c.im]
re (ci)	interval	Real part of ci
im (ci)	interval	Imaginary part of ci
inf (ci)	complex	Complex lower bound z of ci with $z = (ci.re.inf, ci.im.inf)$
sup (ci)	complex	Complex upper bound z of ci with $z = (ci.re.sup, ci.im.sup)$

r = *real* expression, $i, i1, i2$ = *interval* expression,
$c, c1, c2$ = *complex* expression, ci = *cinterval* expression

(*)	:	$c1 <= c2$	is assumed, otherwise an error occurs.
(*')	:	$r <= c$	is assumed, otherwise an error occurs.
(*")	:	$c <= r$	is assumed, otherwise an error occurs.

Overloading of the Assignment Operator

The type conversions *real*, *complex*, or *interval* to *cinterval* are provided as overloaded assignment operators:

Assignment	Meaning
ci := r	ci := compl (intval (r))
ci := c	ci := intval (c)
ci := i	ci := compl (i)

ci = *cinterval* variable, i = *interval* expression
c = *complex* expression, r = *real* expression

Predefined Functions

All mathematical functions of PASCAL–XSC available for real arguments are supplied for complex interval arguments *ci*. These complex interval functions F satisfy $F(ci) \supseteq f(ci) = \{f(c) : c \in ci\}$. Moreover, functions for the computation of the angle component of the exponential representation, for the conjugation, and for the computation of midpoint, diameter, and epsilon inflation of a complex interval are available.

Function	Result Type	Meaning	
sqr (ci)	*cinterval*	$(ci)^2$	Square
sqrt (ci)	*cinterval*	$\sqrt{ci}$	Square Root
exp (ci)	*cinterval*	e^{ci}	Exponential Function
exp2 (ci)	*cinterval*	2^{ci}	Power Function, Base 2
exp10 (ci)	*cinterval*	10^{ci}	Power Function, Base 10
ln (ci)	*cinterval*	ln (ci)	Natural Logarithm
log2 (ci)	*cinterval*	$\log_2(ci)$	Logarithm, Base 2
log10 (ci)	*cinterval*	$\log_{10}(ci)$	Logarithm, Base 10
sin (ci)	*cinterval*	sin (ci)	Sine
cos (ci)	*cinterval*	cos (ci)	Cosine
tan (ci)	*cinterval*	tan (ci)	Tangent
cot (ci)	*cinterval*	cot (ci)	Cotangent
arcsin (ci)	*cinterval*	arcsin (ci)	Arc Sine
arccos (ci)	*cinterval*	arccos (ci)	Arc Cosine
arctan (ci)	*cinterval*	arctan (ci)	Arc Tangent
arccot (ci)	*cinterval*	arccot (ci)	Arc Cotangent

ci = *cinterval* expression

Function	Result Type		Meaning
sinh (ci)	cinterval	sinh (ci)	Hyperbolic Sine
cosh (ci)	cinterval	cosh (ci)	Hyperbolic Cosine
tanh (ci)	cinterval	tanh (ci)	Hyperbolic Tangent
coth (ci)	cinterval	coth (ci)	Hyperbolic Cotangent
arsinh (ci)	cinterval	arsinh (ci)	Inverse Hyperbolic Sine
arcosh (ci)	cinterval	arcosh (ci)	Inverse Hyperbolic Cosine
artanh (ci)	cinterval	artanh (ci)	Inverse Hyperbolic Tangent
arcoth (ci)	cinterval	arcoth (ci)	Inverse Hyperb. Cotangent
conj (ci)	cinterval	$\overline{ci} = a - ib$	Conjugation of $ci = a + ib$
abs (ci)	interval	$j = \sqrt{ci.re^2 + ci.im^2}$	Absolute Value of ci
arg (ci)	interval	φ	Angle component of the exponential representation of ci
mid (ci)	complex	m	Midpoint of ci
diam (ci)	real	d	Diameter of ci
blow (ci,r)	cinterval	‡	Epsilon Inflation

$ci = cinterval$ expression,   r = $real$ expression

‡ :   blow := compl (blow(ci.re,r),blow(ci.im,r))

The domains and ranges of the predefined functions are implementation-dependent and are described in the user manual.

Example 3.3.2:

If a of type *cinterval* is defined as

a := compl (intval $(-1,3)$, intval $(3,4)$),

then the functions *abs* and *sqr* deliver the results:

Expression	Result
abs (a)	$[3,5]$
sqr (a)	$[-16,0] + i\,[-8,24]$

Input/Output Procedures

This module supplies the procedures

procedure read (**var** f: text; **var** a: cinterval);
procedure write (**var** f: text; a: cinterval);

with optional file parameters, arbitrarily many input/output parameters, but without format specifications.

A complex interval $ci = [x, y] + i[v, w]$ must be entered in the form

$([x, y], [v, w])$ general complex interval

or in the form

$(x, [v, w])$ with $x = y$

or in the form

$([x, y], v)$ with $v = w$

or in the form

$[x, y]$ with $v = w = 0$

or in the form

(x, v) with $x = y$ and $v = w$

or in the form

x with $x = y$ and $v = w = 0$.

The rounding of real and imaginary part is done as described in section 3.2.

The output of a complex interval is done with the interval rounding described in section 3.2 for the real and imaginary parts. The result has the form

$([x, y], [v, w])$,

with an implementation-dependent default format for the *real* values x, y, v, and w.

Example 3.3.3:

If *ci1*, *ci2*, and *ci3* are of type *cinterval*, then the statements

        read (ci1, ci2, ci3);
        writeln (ci1);
        writeln (ci2);
        writeln (ci3);

accept the input data

        [4,5]
        (8,10)
        100

and write the complex intervals

        ([4.0E+00, 5.0E+00],[0.0E+00, 0.0E+00])
        ([8.0E+00, 8.0E+00],[1.0E+01, 1.0E+01])
        ([1.0E+02, 1.0E+02],[0.0E+00, 0.0E+00])

Another *real* representation may be used depending on the implementation.

3.4   The Module MV_ARI

Real Matrix/Vector Arithmetic

This module supplies all operators, functions, and procedures necessary for computations with real vectors and matrices.

Types

The dynamic types for representing real vectors and matrices defined by

> **type** rvector = **dynamic array** [*] **of** real;
> rmatrix= **dynamic array** [*] **of** rvector;

are part of the language core of PASCAL–XSC. The actual index bounds are specified in connection with the declaration of variables of these types.

Operators

Many of the basic matrix/vector operations known from mathematics are predefined
in this module. There are the monadic operators $+, -$ and the four basic operations
$+, -, *, /$, each with three different kinds of rounding. Special combinations of mixed
types of operands are permitted. The operations $+$ and $-$ for vectors and matrices
are defined componentwise by

$$
\begin{array}{llllllll}
c & := & a & \pm & b & \text{with} & c[i] & := & a[i] \pm b[i] \\
C & := & A & \pm & B & \text{with} & C[i,j] & := & A[i,j] \pm B[i,j]
\end{array}
$$

with a, b, c of type $rvector$, and A, B, C of type $rmatrix$. The operators $*$ and $/$
are defined by

$$
\begin{array}{llllllll}
s & := & a & * & b & \text{with} & s & := & \#* \ (\textbf{for } i:=\text{lbound}(a) \textbf{ to } \text{ubound}(a) \\
& & & & & & & & \qquad \text{sum} \ (a[i]*b[i])) \quad \ddagger
\end{array}
$$

$$
\begin{array}{llllllll}
c & := & r & * & a & \text{with} & c[i] & := & r * a[i] \\
c & := & a & * & r & \text{with} & c[i] & := & a[i] * r \\
c & := & a & / & r & \text{with} & c[i] & := & a[i] / r \\
c & := & A & * & b & \text{with} & c[i] & := & A[i] * b \quad \ddagger \\
C & := & r & * & A & \text{with} & C[i,j] & := & r * A[i,j] \\
C & := & A & * & r & \text{with} & C[i,j] & := & A[i,j] * r \\
C & := & A & / & r & \text{with} & C[i,j] & := & A[i,j] / r \\
C & := & A & * & B & \text{with} & C[i,j] & := & A[i] * B[*,j] \quad \ddagger
\end{array}
$$

$\ddagger$: Scalar product
with maximum accuracy

with r, s of type $real$, a, b, c of type $rvector$ and A, B, C of type $rmatrix$. The
operations with directed rounding are defined in a corresponding way.

The definition of the relational operators $=, <>, <, <=, >, >=$ is based upon $=$ and $<=$. It is realized according to (RD) (see page 130). If a and b are of type $rvector$, and A and B are of type $rmatrix$, then

$$
\begin{array}{llll}
a & <= & b & \Longleftrightarrow & a[i] & <= & b[i] & \text{for all } i \\
A & <= & B & \Longleftrightarrow & A[i,j] & <= & B[i,j] & \text{for all } i, j
\end{array}
$$

The operators on the right side of the equivalences are the ones for values of type $real$.

left Operand \ right Operand	integer real	rvector	rmatrix
monadic		$+, -$	$+, -$
integer real		$*, *<, *>$	$*, *<, *>$
rvector	$*, *<, *>$ $/, /<, />$	$\circ$ $\vee$	
rmatrix	$*, *<, *>$ $/, /<, />$	$*, *<, *>$	$\circ$ $\vee$

The Operators of Module MV_ARI

$$\circ \in \{+, +<, +>, -, -<, ->, *, *<, *>\}$$

$$\vee \in \{=, <>, <, <=, >, >=\}$$

Example 3.4.1:

A Runge-Kutta method can be used for the approximate solution of initial value problems of the form

$$Y' = F(x, Y); \quad Y(x^0) = Y^0;$$

with

$$Y = \begin{pmatrix} y_1(x) \\ \vdots \\ y_n(x) \end{pmatrix}, \quad Y' = \begin{pmatrix} y_1'(x) \\ \vdots \\ y_n'(x) \end{pmatrix}$$

and

$$F(x, Y) = \begin{pmatrix} f_1(x, y_1, \ldots, y_n) \\ \vdots \\ f_n(x, y_1, \ldots, y_n) \end{pmatrix}.$$

To determine an approximation of the solution Y at $x + h$, the formulas

$$K_1 = h * F(x, Y)$$

$$K_2 = h * F(x + \frac{h}{2}, Y + \frac{K_1}{2})$$

$$K_3 = h * F(x + \frac{h}{2}, Y + \frac{K_2}{2})$$

$$K_4 = h * F(x + h, Y + K_3)$$

and

$$Y(x + h) = Y(x) + (K_1 + 2K_2 + 2K_3 + K_4)/6$$

are applied. After the definition of the *rvector* function F and the declaration of the variables *k1, k2, k3, k4, Y* of type *rvector* and *h, x* of type *real*, these formulas can directly be transferred into PASCAL–XSC source code:

```
k1  :=  h * F (x , Y);
k2  :=  h * F (x + h/2, Y + k1/2);
k3  :=  h * F (x + h/2, Y + k2/2);
k4  :=  h * F (x + h , Y + k3);
Y   :=  Y + (k1 + 2 * k2 + 2 * k3 + k4) / 6;
```

Overloading of the Assignment Operator

The componentwise initialization of *rvector* and *rmatrix* variables is provided as overloaded assignment operators:

Assignment		Meaning
rv := r	rv[j] := r	$j = \text{lb(rv)},...,\text{ub(rv)}$
rM := r	rM[j,k] := r	$j = \text{lb(rM,1)},...,\text{ub(rM,1)}$ $k = \text{lb(rM,2)},...,\text{ub(rM,2)}$

r = *real* expression, rv = *rvector* variable, rM = *rmatrix* variable

Predefined Functions

The functions *id* and *null* are supplied for generating an identity matrix and a null matrix or a null vector. Furthermore, the function *transp* computes the transposed of a matrix.

Function	Result Type	Meaning
null (v)	*rvector*	Null vector with the actual index range of v
vnull (n)	*rvector*	Null vector with the index range [1..n]
null (M)	*rmatrix*	Null matrix with the actual index ranges of M
null (M1,M2)	*rmatrix*	Null matrix with the actual index ranges of the product matrix M1 · M2
null (n)	*rmatrix*	Null matrix with index range [1..n,1..n]
null (n1,n2)	*rmatrix*	Null matrix with index range [1..n1,1..n2]
id (M)	*rmatrix*	Identity matrix with the actual index ranges of M
id (M1,M2)	*rmatrix*	Identity matrix with the actual index ranges of the product matrix M1 · M2
id (n)	*rmatrix*	Identity matrix with the index ranges [1..n,1..n]
id (n1,n2)	*rmatrix*	Identity matrix with the index ranges [1..n1,1..n2]
transp (M)	*rmatrix*	Transposed matrix Mt of M with $Mt[i,j] = M[j,i]$

n, n1, n2 = *integer* expression,   v = *rvector* expression
M, M1, M2 = *rmatrix* expression

Example 3.4.2:

If E denotes the identity matrix and $R \approx A^{-1}$ an approximate inverse of the square matrix A, then the defect

$$D = E - R \cdot A$$

is often used in the numerical computations. In PASCAL–XSC, the defect can be computed by

$$D := id (A) - R * A,$$

or with the use of an accurate expression (see section 2.4.4) by

$$D := \#* (id (A) - R * A) .$$

In the second form, the defect matrix is computed with only one rounding in each component.

Input/Output Procedures

The procedures

> **procedure** read (**var** f: text; **var** a: rvector);
> **procedure** read (**var** f: text; **var** A: rmatrix);
> **procedure** write (**var** f: text; a: rvector);
> **procedure** write (**var** f: text; A: rmatrix);

are provided with optional file parameters, arbitrarily many input/output parameters, but without any format specifications.

The input of a vector or a matrix is done componentwise according to the input of *real* values. A matrix is entered row by row. The output of a vector or a matrix is also done componentwise using an implementation-dependent default format for the real components.

Example 3.4.3:

The statement

> read (b, A, x)

reads the vector b, the matrix A and the vector x.

3.5   The Module MVC_ARI

Complex Matrix/Vector Arithmetic

This module supplies all operators, functions, and procedures necessary for computations with complex vectors and matrices.

Types

The dynamic types for representing complex vectors and matrices defined by

> type cvector = **dynamic array** [*] of complex;
> cmatrix= **dynamic array** [*] of cvector;

are part of the language core of PASCAL–XSC. The actual index bounds are specified in connection with the declaration of variables of these types.

Operators

Many of the basic complex matrix/vector operations known from mathematics are
predefined in this module. There are the monadic operators $+, -$ and the four
basic operations $+, -, *, /$, each with three different kinds of rounding. Special
combinations of mixed types of operands are permitted. The operations $+$ and $-$
for complex vectors and matrices are defined componentwise by

$$c := a \pm b \quad \text{with} \quad c[i] := a[i] \pm b[i]$$
$$C := A \pm B \quad \text{with} \quad C[i,j] := A[i,j] \pm B[i,j]$$

with a, b, and c of type cvector, and A, B, and C of type cmatrix. The operators
$*$ and $/$ are defined by

$$s := a * b \quad \text{with} \quad s := \#* \text{ (for i:=lbound(a) to ubound(a)}$$
$$\text{sum } (a[i]*b[i])) \quad \ddagger$$

$$c := r * a \quad \text{with} \quad c[i] := r * a[i]$$
$$c := a * r \quad \text{with} \quad c[i] := a[i] * r$$
$$c := a / r \quad \text{with} \quad c[i] := a[i] / r$$
$$c := A * b \quad \text{with} \quad c[i] := A[i] * b \quad \ddagger$$
$$C := r * A \quad \text{with} \quad C[i,j] := r * A[i,j]$$
$$C := A * r \quad \text{with} \quad C[i,j] := A[i,j] * r$$
$$C := A / r \quad \text{with} \quad C[i,j] := A[i,j] / r$$
$$C := A * B \quad \text{with} \quad C[i,j] := A[i] * B[*,j] \quad \ddagger$$

> $\ddagger$:   Scalar product
> with maximum accuracy

with r and s of type complex, a, b. and c of type cvector, and A, B, and C of
type cmatrix. The operations with mixed operand types and the operations with
directed rounding are defined in a corresponding way.

The definition of the relational operators $=, <>, <, <=, >, >=$ is based upon $=$ and $<=$. It is realized according to (RD) (see page 130). If a and b are of type *cvector* and A and B are of type *cmatrix*, then

$$a \ <= \ b \ \Longleftrightarrow \ a[i] \ <= \ b[i] \quad \text{for all } i$$
$$A \ <= \ B \ \Longleftrightarrow \ A[i,j] \ <= \ B[i,j] \quad \text{for all } i, j$$

The operators on the right side of the equivalences are the ones for values of type *complex*.

left Operand \ right Operand	integer real	complex	rvector	cvector	rmatrix	cmatrix
monadic				+,−		+,−
integer real				*, *<, *>		*, *<, *>
complex			*, *<, *>	*, *<, *>	*, *<, *>	*, *<, *>
rvector		*, *<, *>, /, /<, />		o / v		
cvector	*, *<, *>, /, /<, />	*, *<, *>, /, /<, />	o / v	o / v		
rmatrix		*, *<, *>, /, /<, />		*, *<, *>		o / v
cmatrix	*, *<, *>, /, /<, />	*, *<, *>, /, /<, />	*, *<, *>	*, *<, *>	o / v	o / v

The Operators of Module MVC_ARI

$$\circ \in \{+, +<, +>, -, -<, ->, *, *<, *>\}$$

$$\triangledown \in \{=, <>, <, <=, >, >=\}$$

Example 3.5.1:

If *cv* is of type *cvector* and *cM* is of type *cmatrix*, then a stretching with the factor 1/3 may be produced by

cv := cv / 3;
cM := cM / 3;

The division operation may also be executed with downwardly or upwardly rounding using /< or />, respectively.

Transfer Functions for Complex Vectors

The following transfer functions are supplied for type conversion between the types *rvector* and *cvector*:

Function	Result Type	Meaning
compl (rv1,rv2)	cvector	Complex vector cv with $cv[i] = compl\ (rv1[i],\ rv2[i])$
compl (rv)	cvector	Complex vector cv with $cv[i] = compl\ (rv[i])$
re (cv)	rvector	Real part vector rv with $rv[i] = re\ (cv[i])$
im (cv)	rvector	Imaginary part vector rv with $rv[i] = im\ (cv[i])$

rv, rv1, rv2 = *rvector* expression,   cv = *cvector* expression

Transfer Functions for Complex Matrices

The following transfer functions are supplied for type conversion between the types *rmatrix* and *cmatrix*:

Function	Result Type	Meaning
compl (rM1,rM2)	cmatrix	Complex matrix cM with $cM[i,j] = compl\ (rM1[i,j],\ rM2[i,j])$
compl (rM)	cmatrix	Complex matrix cM with $cM[i,j] = compl\ (rM[i,j])$
re (cM)	rmatrix	Real part matrix rM with $rM[i,j] = re\ (cM[i,j])$
im (cM)	rmatrix	Imaginary part matrix rM with $rM[i,j] = im\ (cM[i,j])$

rM, rM1, rM2 = *rmatrix* expression,   cM = *cmatrix* expression

Overloading of the Assignment Operator

The componentwise initialization of *cvector* and *cmatrix* variables and type conversions from *rvector* to *cvector* and *rmatrix* to *cmatrix* are provided as overloaded assignment operators:

Assignment	Meaning		
cv := r	cv[j] :=	compl (r)	$j = lb(cv),...,ub(cv)$
cv := c	cv[j] :=	c	$j = lb(cv),...,ub(cv)$
cv := rv	cv :=	compl (rv)	
cM := r	cM[j,k] :=	compl (r)	$j = lb(cM,1),...,ub(cM,1)$ $k = lb(cM,2),...,ub(cM,2)$
cM := c	cM[j,k] :=	c	$j = lb(cM,1),...,ub(cM,1)$ $k = lb(cM,2),...,ub(cM,2)$
cM := rM	cM :=	compl (rM)	

c = *complex* expression,   cv = *cvector* variable
cM = *cmatrix* variable,   r = *real* expression
rv = *rvector* expression,   rM = *rmatrix* expression

Predefined Functions

The functions *id* and *null* are supplied for generating an identity matrix and a null matrix or a null vector. The functions *transp* and *herm* compute the transposed and the Hermitian matrices. The function *conj* for conjugation is available, too.

Function	Result Type	Meaning
null (cv)	*rvector*	Null vector with the actual index range of *cv*
null (cM)	*rmatrix*	Null matrix with the actual index ranges of *cM*
null (cM1,cM2)	*rmatrix*	Null matrix with the actual index ranges of the product matrix *cM1 · cM2*
id (cM)	*rmatrix*	Identity matrix with the actual index ranges of *cM*
id (cM1,cM2)	*rmatrix*	Identity matrix with the actual index ranges of the product matrix *cM1 · cM2*

cv = *cvector* expression,   cM, cM1, cM2 = *cmatrix* expression

Function	Result Type	Meaning
conj (cv)	cvector	Conjugated complex vector cvc with $cvc[i] = conj\ (cv[i])$
conj (cM)	cmatrix	Conjugated complex matrix cMc with $cMc[i,j] = conj\ (cM[i,j])$
transp (cM)	cmatrix	Transposed matrix cMt of cM with $cMt[i,j] = cM[j,i]$
herm (cM)	cmatrix	Hermitian matrix cMh of cM with $cMh[i,j] = conj\ (cM[j,i])$

cv, cvc = cvector expression,   cM, cMc, cMt, cMh = cmatrix expression

Example 3.5.2:

If cM, $cM1$, and $cM2$ are complex matrices of type cmatrix, and the statements

    cM1 := conj (transp (cM));
    cM2 := herm (cM);

are executed, then the boolean expression

    cM1 = cM2

delivers *true*.

Input/Output Procedures

The procedures

    procedure read (var f: text; var a: cvector);
    procedure read (var f: text; var A: cmatrix);
    procedure write (var f: text; a: cvector);
    procedure write (var f: text; A: cmatrix);

are provided with optional file parameters, arbitrary many input/output parameters, but without any format specifications.

A complex vector or a complex matrix is entered componentwise according to the input of *complex* values. A matrix is entered row by row. The output of a complex vector or a complex matrix is also done componentwise using the default output format for complex numbers.

3.6   The Module MVI_ARI

Interval Matrix/Vector Arithmetic

This module supplies all operators, functions, and procedures necessary for computations with interval vectors and matrices.

Types

The dynamic types for representing interval vectors and matrices defined by

> **type** ivector = **dynamic array** [∗] **of** interval;
>   imatrix= **dynamic array** [∗] **of** ivector;

are part of the language core of PASCAL–XSC. The actual index bounds are specified in connection with the declaration of variables of these types.

Operators

Many of the basic interval matrix/vector operations known from the mathematics are predefined in this module. There are the monadic operators $+, -$ and the four basic operations $+, -, *, /$, each with componentwise rounding to the smallest enclosing interval. Even special combinations of mixed types of operands are permitted. The operations $+$ and $-$ for interval vectors and matrices are defined componentwise by

$$c := a \pm b \quad \text{with} \quad c[i] := a[i] \pm b[i]$$
$$C := A \pm B \quad \text{with} \quad C[i,j] := A[i,j] \pm B[i,j]$$

with a, b, and c of type *ivector* and A, B, and C of type *imatrix*. The operators $*$ and $/$ are defined by

$$s := a * b \quad \text{with} \quad s := \#\# \left(\textbf{for } i:=\text{lbound}(a) \textbf{ to } \text{ubound}(a) \right.$$
$$\left.\textbf{sum } (a[i]*b[i])\right) \quad \ddagger$$
$$c := r * a \quad \text{with} \quad c[i] := r * a[i]$$
$$c := a * r \quad \text{with} \quad c[i] := a[i] * r$$
$$c := a / r \quad \text{with} \quad c[i] := a[i] / r$$
$$c := A * b \quad \text{with} \quad c[i] := A[i] * b \quad \ddagger$$
$$C := r * A \quad \text{with} \quad C[i,j] := r * A[i,j]$$
$$C := A * r \quad \text{with} \quad C[i,j] := A[i,j] * r$$
$$C := A / r \quad \text{with} \quad C[i,j] := A[i,j] / r$$
$$C := A * B \quad \text{with} \quad C[i,j] := A[i] * B[*,j] \quad \ddagger$$

$\ddagger$:   Scalar product
      with maximum accuracy

with r and s of type *interval*, a, b, and c of type *ivector*, and A, B, and C of type *imatrix*. The operations with mixed operand types are defined in a corresponding way.

The definition of the relational operators $=, <>, <, <=, >, >=$ is based upon $=$ and $<=$. It is realized according to (RD) (see page 130). If a and b are of type *ivector*, and A and B of type *imatrix*, then

$$a \quad <= \quad b \quad \Longleftrightarrow \quad a[i] \quad <= \quad b[i] \quad \text{for all } i$$
$$A \quad <= \quad B \quad \Longleftrightarrow \quad A[i,j] \quad <= \quad B[i,j] \quad \text{for all } i, j$$

The operators on the right side of the equivalences are the ones for values of type *interval*.

Moreover, the operators **in** for the relations "is contained in" and "is contained in the interior" and the operator $><$ to test for disjointedness are provided for interval vectors and interval matrices. These operators are defined componentwise.

The lattice operators $+*$ and $**$ denote the interval hull and the interval intersection as described for the type *interval* in section 3.2 (I_ARI).

left Operand \ right Operand	integer real	interval	rvector	ivector	rmatrix	imatrix
monadic				$+, -$		$+, -$
integer real				$*$		$*$
interval			$*$	$*$	$*$	$*$
rvector		$*, /$	$+*$	◇ $=, <>, \text{in}$ $+*$		
ivector	$*, /$	$*, /$	◇ $=, <>$ $+*$	◇ $\text{in}, \vee, ><$ $+*, **$		
rmatrix		$*, /$		$*$	$+*$	◇ $=, <>, \text{in}$ $+*$
imatrix	$*, /$	$*, /$	$*$	$*$	◇ $=, <>$ $+*$	◇ $\text{in}, \vee, ><$ $+*, **$

The Operators of Module MVI_ARI

$$\diamond \in \{+, -, *\}$$

$$\vee \in \{=, <>, <, <=, >, >=\}$$

Example 3.6.1:

If we want to execute the Runge-Kutta method mentioned in section 3.4 (MV_ARI) using interval arithmetic, we only have to change a few things in the original program. The variables *k1, k2, k3, k4, Y* must be declared of type *ivector*, and the function *F* must be defined with result type *ivector*. If we now use the module MVI_ARI instead of MV_ARI, the program statements

```
k1  :=  h * F (x , Y);
k2  :=  h * F (x + h/2, Y + k1/2);
k3  :=  h * F (x + h/2, Y + k2/2);
k4  :=  h * F (x + h , Y + k3);
Y   :=  Y + (k1 + 2 * k2 + 2 * k3 + k4) / 6;
```

need not be changed in any way. Now, all operators denote the corresponding interval operations. Although we have an interval result *Y*, we *do not* have an enclosure for the true solution of the ordinary differential equation. To compute an enclosure, we would also have to enclose the truncation error of the method.

Transfer Functions for Interval Vectors

The following transfer functions are supplied for type conversions between the types *rvector* and *ivector*:

Function	Result Type	Meaning
intval (rv1,rv2)	ivector	Interval vector *iv* with $iv[i] = intval\ (rv1[i],rv2[i])$ [*]
intval (rv)	ivector	Interval vector *iv* with $iv[i] = intval\ (rv[i])$
inf (iv)	rvector	Vector *rv* of lower bounds with $rv[i] = inf\ (iv[i])$
sup (iv)	rvector	Vector *rv* of upper bounds with $rv[i] = sup\ (iv[i])$

rv, rv1, rv2 = *rvector* expression,   iv = *ivector* expression
[*]: rv1 <= rv2 is assumed, otherwise an error occurs.

Transfer Functions for Interval Matrices

The following transfer functions are supplied for type conversions between the types *rmatrix* and *imatrix*:

Function	Result Type	Meaning
intval (rM1,rM2)	*imatrix*	Interval vector *iM* with $iM[i,j] = intval\ (rM1[i,j],rM2[i,j])$ (*)
intval (rM)	*imatrix*	Interval vector *iM* with $iM[i,j] = intval\ (rM[i,j])$
inf (iM)	*rmatrix*	Vector *rM* of lower bounds with $rM[i,j] = inf\ (iM[i,j])$
sup (iM)	*rmatrix*	Vector *rM* of upper bounds with $rM[i,j] = sup\ (iM[i,j])$

rM, rM1, rM2 = rmatrix expression,   iM = imatrix expression
(*): rM1 <= rM2 is assumed, otherwise an error occurs.

Overloading of the Assignment Operator

The componentwise initialization of *ivector* and *imatrix* variables and type conversions from *rvector* to *ivector* and *rmatrix* to *imatrix* are provided as overloaded assignment operators:

Assignment		Meaning		
iv := r	iv[j]	:=	intval (r)	j = lb(iv),...,ub(iv)
iv := i	iv[j]	:=	i	j = lb(iv),...,ub(iv)
iv := rv	iv	:=	intval (rv)	
iM := r	iM[j,k]	:=	intval (r)	j = lb(iM,1),...,ub(iM,1) k = lb(iM,2),...,ub(iM,2)
iM := i	iM[j,k]	:=	i	j = lb(iM,1),...,ub(iM,1) k = lb(iM,2),...,ub(iM,2)
iM := rM	iM	:=	intval (rM)	

i = interval expression, iv = ivector variable, iM = imatrix variable
r = real expression, rv = rvector expression, rM = rmatrix expression

Predefined Functions

The functions *id* and *null* are supplied for generating an identity matrix and a null matrix or a null vector. The function *transp* computes the transposed matrix. The functions *mid*, *diam*, and *blow* compute midpoint, diameter, and epsilon inflation componentwise.

Function	Result Type	Meaning
null (iv)	rvector	Null vector with the actual index range of *iv*
null (iM)	rmatrix	Null matrix with the actual index ranges of *iM*
null (iM1,iM2)	rmatrix	Null matrix with the actual index ranges of the product matrix *iM1 · iM2*
id (iM)	rmatrix	Identity matrix with the actual index ranges of *iM*
id (iM1,iM2)	rmatrix	Identity matrix with the actual index ranges of the product matrix *iM1 · iM2*
mid (iv)	rvector	Midpoint vector *rv* with $rv[i] = mid\ (iv[i])$
diam (iv)	rvector	Diameter vector *rv* with $rv[i] = diam\ (iv[i])$
mid (iM)	rmatrix	Midpoint matrix *rM* with $rM[i,j] = mid\ (iM[i,j])$
diam (iM)	rmatrix	Diameter matrix *rM* with $rM[i,j] = diam\ (iM[i,j])$
transp (iM)	imatrix	Transposed matrix *iMt* of *iM* with $iMt[i,j] = iM[j,i]$
blow (iv,r)	ivector	Vector epsilon inflation *ive* with $ive[i] = blow\ (iv[i],r)$
blow (iM,r)	imatrix	Matrix epsilon inflation *iMe* with $iMe[i,j] = blow\ (iM[i,j],r)$

r = *real* expression   iv, ive = *ivector* expression
iM, iM1, iM2, iMt, iMe = *imatrix* expression

Example 3.6.2:

An interval enclosure for the defect

$$D = E - R \cdot A$$

mentioned in section 3.4 (MV_ARI) can be computed by using MVI_ARI, the variables *A*, *D*, and *R* of type *imatrix*, and the statement

$$D := id\ (A) - R * A.$$

The tightest possible enclosing interval matrix can be computed using an accurate expression (see section 2.4.4)

$$D := \#\# \, (\, \text{id} \, (A) - R * A \,).$$

Input/Output Procedures

The procedures

procedure read (**var** f: text; **var** a: ivector);
procedure read (**var** f: text; **var** A: imatrix);
procedure write (**var** f: text; a: ivector);
procedure write (**var** f: text; A: imatrix);

are provided with optional file parameters, arbitrarily many input/output parameters, but without any format specifications.

An interval vector or an interval matrix is entered componentwise as individual *interval* values. A matrix is entered row by row. The output of an interval vector or an interval matrix is also done componentwise using the default output format for intervals.

3.7 The Module MVCI_ARI

Complex Interval Matrix/Vector Arithmetic

This module supplies all operators, functions, and procedures necessary for computations with complex interval vectors and matrices.

Types

The dynamic types for representing complex interval vectors and matrices defined by

> type civector = **dynamic array** [*] **of** cinterval;
> cimatrix= **dynamic array** [*] **of** civector;

are part of the language core of PASCAL–XSC. The actual index bounds are specified in connection with the declaration of variables of these types.

Operators

Many of the basic complex interval matrix/vector operations known from the mathematics are predefined in this module. There are the monadic operators $+, -$ and the four basic operations $+, -, *, /$, each with componentwise rounding to the smallest enclosing complex interval. Even special combinations of mixed types of operands are permitted. The operations $+$ and $-$ for complex interval vectors and matrices are defined componentwise by

$$
\begin{array}{lllllll}
c & := & a & \pm & b & \text{with} & c[i] & := & a[i] \pm b[i] \\
C & := & A & \pm & B & \text{with} & C[i,j] & := & A[i,j] \pm B[i,j]
\end{array}
$$

with a, b, and c of type *civector* and A, B, and C of type *cimatrix*. The operators $*$ and $/$ are defined by

$$
\begin{array}{lllllll}
s & := & a & * & b & \text{with} & s & := & \#\# \ (\text{for } i:=\text{lbound}(a) \ \text{to ubound}(a) \\
& & & & & & & & \quad \text{sum } (a[i]*b[i])) \quad \ddagger
\end{array}
$$

$$
\begin{array}{lllllll}
c & := & r & * & a & \text{with} & c[i] & := & r * a[i] \\
c & := & a & * & r & \text{with} & c[i] & := & a[i] * r \\
c & := & a & / & r & \text{with} & c[i] & := & a[i] / r \\
c & := & A & * & b & \text{with} & c[i] & := & A[i] * b \quad \ddagger \\
C & := & r & * & A & \text{with} & C[i,j] & := & r * A[i,j] \\
C & := & A & * & r & \text{with} & C[i,j] & := & A[i,j] * r \\
C & := & A & / & r & \text{with} & C[i,j] & := & A[i,j] / r \\
C & := & A & * & B & \text{with} & C[i,j] & := & A[i] * B[*,j] \quad \ddagger
\end{array}
$$

> $\ddagger$:   Scalar product
> with maximum accuracy

with r and s of type *cinterval*, a, b, and c of type *civector*, and A, B, and C of type *cimatrix*. The operations with mixed operand types are defined in a corresponding way.

The definition of the relational operators $=, <>, <, <=, >, >=$ is based upon $=$ and $<=$. It is realized according to (RD) (see page 130). If a and b are of type *civector* and A and B are of type *cimatrix*, then

$$
\begin{aligned}
a &<= b &\Longleftrightarrow\quad a[i] &<= b[i] &\text{for all } i \\
A &<= B &\Longleftrightarrow\quad A[i,j] &<= B[i,j] &\text{for all } i, j
\end{aligned}
$$

The operators on the right side of the equivalences are the ones for complex intervals.

Moreover, the operators **in** for the relations "is contained in" and "is contained in the interior", and the operator $><$ for the test on disjointedness are provided for complex interval vectors and complex interval matrices. These operators are defined componentwise.

The lattice operators $+*$ and $**$ denote the complex interval hull and the complex interval intersection as described for the type *cinterval* in section 3.3 (CI_ARI).

The review of the operators defined in Module MVCI_ARI is given in two tables due to the large number of operators. The first table has no matrix types as right operands, while the second table has *only* matrix types as right operands.

left Operand \ right Operand	integer real	complex	interval	cinterval	rvector	cvector	ivector	civector
monadic								+, −
integer real								*
complex							*	*
interval						*		*
cinterval					*	*	*	*
rvector				*,/		+*		◇ =,<>,in +*
cvector			*,/	*,/	+*	+*	◇ =,<>,in +*	◇ =,<>,in +*
ivector		*,/		*,/		◇, =,<>, +*		◇ in,∨,><, +*,**
civector	*,/	*,/	*,/	*,/	◇, =,<>, +*	◇, =,<>, +*	◇ ∨,><, +*,**	◇ in,∨,><, +*,**
rmatrix				*,/				*
cmatrix			*,/	*,/			*	*
imatrix		*,/		*,/		*		*
cimatrix	*,/	*,/	*,/	*,/	*	*	*	*

The Operators of Module MVCI_ARI (Part 1)

$$\diamond \in \{+, -, *\}$$

$$\vee \in \{=, <>, <, <=, >, >=\}$$

left Operand \ right Operand	rmatrix	cmatrix	imatrix	cimatrix
monadic				+, −
integer real				*
complex			*	*
interval		*		*
cinterval	*	*	*	*
rvector				
cvector				
ivector				
civector				
rmatrix		+*		◇ =, <>, in +*
cmatrix	+*	+*	◇ =, <>, in +*	◇ =, <>, in +*
imatrix		◇, =, <>, +*		◇ in, V, ><, +*, **
cimatrix	◇, =, <>, +*	◇, =, <>, +*	◇ V, ><, +*, **	◇ in, V, ><, +*, **

The Operators of Module MVCI_ARI (Part 2)

$$\diamond \in \{+, -, *\}$$

$$\mathsf{V} \in \{=, <>, <, <=, >, >=\}$$

Transfer Functions for Complex Interval Vectors

The following transfer functions are supplied for type conversions between the types *rvector*, *cvector*, *ivector*, and *civector*:

Function	Result Type	Meaning
compl (iv1,iv2)	civector	Complex interval vector *civ* with $civ[i] = compl\ (iv1[i],iv2[i])$
compl (rv,iv)	civector	Complex interval vector *civ* with $civ[i] = compl\ (rv[i],iv[i])$
compl (iv,rv)	civector	Complex interval vector *civ* with $civ[i] = compl\ (iv[i],rv[i])$
compl (iv)	civector	Complex interval vector *civ* with $civ[i] = compl\ (iv[i])$
intval (cv1,cv2)	civector	Complex interval vector *civ* with $civ[i] = intval\ (cv1[i],cv2[i])$
intval (rv,cv)	civector	Complex interval vector *civ* with $civ[i] = intval\ (rv[i],cv[i])$
intval (cv,rv)	civector	Complex interval vector *civ* with $civ[i] = intval\ (cv[i],rv[i])$
intval (cv)	civector	Interval vector *civ* with $civ[i] = intval\ (cv[i])$
re (civ)	ivector	Real part vector *iv* with $iv[i] = re\ (civ[i])$
im (civ)	ivector	Imaginary part vector *iv* with $iv[i] = im\ (civ[i])$
inf (civ)	cvector	Complex vector *cv* of the lower bounds with $cv[i] = inf\ (civ[i])$
sup (civ)	cvector	Complex vector *cv* of the upper bounds with $cv[i] = sup\ (civ[i])$

rv = *rvector* expression,   cv, cv1, cv2 = *cvector* expression,
iv, iv1, iv2 = *ivector* expression,   civ = *civector* expression

Transfer Functions for Complex Interval Matrices

The following transfer functions are supplied for type conversions between the types
rmatrix, *cmatrix*, *imatrix*, and *cimatrix*:

Function	Result Type	Meaning
compl (iM1,iM2)	cimatrix	Complex interval matrix ciM with $ciM[i,j] = compl\ (iM1[i,j],iM2[i,j])$
compl (rM,iM)	cimatrix	Complex interval matrix ciM with $ciM[i,j] = compl\ (rM[i,j],iM[i,j])$
compl (iM,rM)	cimatrix	Complex interval matrix ciM with $ciM[i,j] = compl\ (iM[i,j],rM[i,j])$
compl (iM)	cimatrix	Complex interval matrix ciM with $ciM[i,j] = compl\ (iM[i,j])$
intval (cM1,cM2)	cimatrix	Complex interval matrix ciM with $ciM[i,j] = intval\ (cM1[i,j],cM2[i,j])$
intval (rM,cM)	cimatrix	Complex interval matrix ciM with $ciM[i,j] = intval\ (rM[i,j],cM[i,j])$
intval (cM,rM)	cimatrix	Complex interval matrix ciM with $ciM[i,j] = intval\ (cM[i,j],rM[i,j])$
intval (cM)	cimatrix	Complex interval matrix ciM with $ciM[i,j] = intval\ (cM[i,j])$
re (ciM)	imatrix	Real part matrix iM with $iM[i,j] = re\ (ciM[i,j])$
im (ciM)	imatrix	Imaginary part matrix iM with $iM[i,j] = im\ (ciM[i,j])$
inf (ciM)	cmatrix	Complex matrix cM of the lower bounds with $cM[i,j] = inf\ (ciM[i,j])$
sup (ciM)	cmatrix	Complex matrix cM of the upper bounds with $cM[i,j] = sup\ (ciM[i,j])$

rM = *rmatrix* expression,   cM, cM1, cM2 = *cmatrix* expression
iM, iM1, iM2 = *imatrix* expression,   ciM = *cimatrix* expression

Overloading of the Assignment Operators

The componentwise initialization of *civector* and *cimatrix* variables and type conversions from *rvector*, *cvector*, or *ivector* to *civector* and *rmatrix*, *cmatrix*, or *imatrix* to *cimatrix* are provided as overloaded assignment operators:

Assignment			Meaning			
civ	:=	r	civ[j]	:=	compl (intval (r))	j = lb(civ),...,ub(civ)
civ	:=	c	civ[j]	:=	intval (c)	j = lb(civ),...,ub(civ)
civ	:=	i	civ[j]	:=	compl (i)	j = lb(civ),...,ub(civ)
civ	:=	ci	civ[j]	:=	ci	j = lb(civ),...,ub(civ)
civ	:=	rv	civ	:=	compl (intval (rv))	
civ	:=	cv	civ	:=	intval (cv)	
civ	:=	iv	civ	:=	compl (iv)	
ciM	:=	r	ciM[j,k]	:=	compl (intval (r))	j = lb(ciM,1),...,ub(ciM,1) k = lb(ciM,2),...,ub(ciM,2)
ciM	:=	c	ciM[j,k]	:=	intval (c)	j = lb(ciM,1),...,ub(ciM,1) k = lb(ciM,2),...,ub(ciM,2)
ciM	:=	i	ciM[j,k]	:=	compl (i)	j = lb(ciM,1),...,ub(ciM,1) k = lb(ciM,2),...,ub(ciM,2)
ciM	:=	ci	ciM[j,k]	:=	ci	j = lb(ciM,1),...,ub(ciM,1) k = lb(ciM,2),...,ub(ciM,2)
ciM	:=	rM	ciM	:=	compl (intval (rM))	
ciM	:=	cM	ciM	:=	intval (cM)	
ciM	:=	iM	ciM	:=	compl (iM)	

ci = *cinterval* expression, civ = *civector* variable, ciM = *cimatrix* variable
i = *interval* expression, iv = *ivector* expression, iM = *imatrix* expression
c = *complex* expression, cv = *cvector* expression, cM = *cmatrix* expression
r = *real* expression, rv = *rvector* expression, rM = *rmatrix* expression

Predefined Functions

The functions *id* and *null* are supplied for generating an identity matrix and a null matrix or a null vector. The functions *transp* and *herm* compute the transposed matrix and the Hermitian matrix. The functions *mid*, *diam*, and *blow* compute the midpoint, diameter, and epsilon inflation componentwise. The function *conj* for the conjugation of complex interval vectors and matrices is also supplied.

Function	Result Type	Meaning
null (civ)	*rvector*	Null vector with the actual index range of *civ*
null (ciM)	*rmatrix*	Null matrix with the actual index ranges of *ciM*
null (ciM1,ciM2)	*rmatrix*	Null matrix with the actual index ranges of the product matrix *ciM1 · ciM2*
id (ciM)	*rmatrix*	Identity matrix with the actual index ranges of *ciM*
id (ciM1,ciM2)	*rmatrix*	Identity matrix with the actual index ranges of the product matrix *ciM1 · ciM2*
mid (civ)	*cvector*	Midpoint vector *cv* with $cv[i] = mid\ (civ[i])$
diam (civ)	*rvector*	Diameter vector *rv* with $rv[i] = diam\ (civ[i])$
mid (ciM)	*cmatrix*	Midpoint matrix *cM* with $cM[i,j] = mid\ (ciM[i,j])$
diam (ciM)	*rmatrix*	Diameter matrix *rM* with $rM[i,j] = diam\ (ciM[i,j])$
conj (civ)	*civector*	Conjugated complex interval vector *civc* with $civc[i] = conj\ (civ[i])$
conj (ciM)	*cimatrix*	Conjugated complex interval matrix *ciMc* with $ciMc[i,j] = conj\ (ciM[i,j])$
transp (ciM)	*cimatrix*	Transposed matrix *ciMt* of *ciM* with $ciMt[i,j] = ciM[j,i]$
herm (ciM)	*cimatrix*	Hermitian matrix *ciMh* of *ciM* with $ciMh[i,j] = conj\ (ciM[j,i])$
blow (civ,r)	*civector*	Vector epsilon inflation *cive* with $cive[i] = blow\ (civ[i],r)$
blow (ciM,r)	*cimatrix*	Matrix epsilon inflation *ciMe* with $ciMe[i,j] = blow\ (ciM[i,j],r)$

n, n1, n2 = *integer* expression,   r = *real* expression,   rv = *rvector* expression
cv = *cvector* expression,   civ, civc, cive = *civector* expression
rM = *rmatrix* expression,   cM = *cmatrix* expression
ciM, ciM1, ciM2, ciMc, ciMt, ciMh, ciMe = *cimatrix* expression

Input/Output Procedures

The Procedures

> **procedure** read (**var** f: text; **var** a: civector);
> **procedure** read (**var** f: text; **var** A: cimatrix);
> **procedure** write (**var** f: text; a: civector);
> **procedure** write (**var** f: text; A: cimatrix);

are provided with optional file parameters, arbitrary many input/output parameters, but without any format specifications.

A complex interval vector or a complex interval matrix is entered componentwise as individual *cinterval* values. A matrix is entered row by row. The output of a complex interval vector or a complex interval matrix is also done componentwise using the default output format for complex intervals.

3.8   The Hierarchy of the Arithmetic Modules

The dependencies between the arithmetic modules generate a hierarchy which is represented in the following diagram. Nevertheless, each arithmetic module that is used in a user-defined program or module must appear in a **use**-clause, since the "lower" modules are linked into the "higher" modules by using a **use**-clause without the reserved word **global**.

For example, a program which uses MVCI_ARI must also use I_ARI when the basic interval operations should be used in the program.

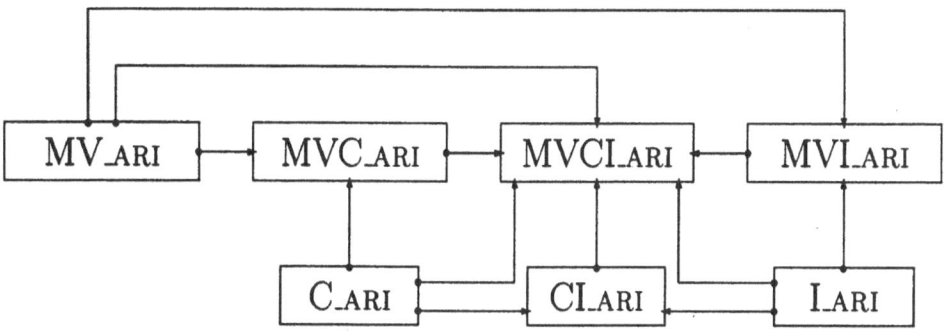

Hierarchy of the PASCAL–XSC Arithmetic Modules

•——▶ stands for " • *is used by* ▶"

3.9   A Complete Sample Program

Here is a complete PASCAL-XSC program which demonstrates the use of some of the arithmetic modules. The module LIN_SOLV is used to enclose the solution of a system of linear equations in an interval vector by successive interval iterations. This module is described later in this section.

The procedure *main*, which is called in the body of *lin_sys*, is only used for reading the dimension of the system and for allocating the dynamic variables. The numerical method itself is started by the call of procedure *linear_system_solver* defined in module LIN_SOLV. This procedure may be called with arrays of arbitrary dimension.

For detailed information on iteration methods with automatic result verification see [46], for example. The use of our program is demonstrated by an example at the end of this section.

The Main Program

program lin_sys (input, output);

```
{ Program for verified solution of linear systems of equations. The   }
{ matrix A and the right-hand side b of the system are to be read in. }
{ The program delivers either a verified solution or an appropriate   }
{ failure message.                                                    }
```

use   lin_solv, { linear system solver   }
              mv_ari,   { matrix/vector arithmetic   }
              mvi_ari; { matrix/vector interval arithmetic }
var
  n : integer;

{- -}

procedure main (n : integer);

```
{ The matrix A and the vectors b and x are allocated dynamically. }
{ The matrix A and the right-hand side b are read in, and         }
{ linear_system_solver is called.                                 }
```

var
  ok : boolean;
  b : rvector[1..n];
  x : ivector[1..n];
  A : rmatrix[1..n,1..n];

begin

  writeln ('Please enter the matrix A:');
  read (A);

  writeln ('Please enter the right-hand side b:');
  read (b);

  linear_system_solver (A, b, x, ok);

  if ok **then**

    begin
      writeln ('The matrix A is non-singular. The solution ');
      writeln ('of the linear system is contained in:');
      write (x);
    end

  else

    writeln ('No solution found !');

end;   {procedure main}

{- -}

begin

  write ('Please enter the dimension n of the linear system: ');
  read (n);
  if n > 0 **then**
    main (n)
  else
    writeln ('Dimension > 0 expected!');

end. {program lin_sys}

The Module LIN_SOLV

```
module lin_solv;

{ Verified solution of the linear system of equations Ax = b. }

use i_ari,    { interval arithmetic                }
    mv_ari,  { matrix/vector arithmetic          }
    mvi_ari; { matrix/vector interval arithmetic }

priority
  inflated_by = *;   { priority level 2 }
```

{- -}

```
operator inflated_by (a : ivector; eps : real) infl: ivector[1..ubound(a)];

{ Computes the epsilon inflation of an interval vector. }

var
  i : integer;
  x : interval;
begin
  for i:= 1 to ubound(a) do
  begin
    x:= a[i];
    if (diam (x) <> 0) then
      a[i] := (1+eps)*x − eps*x
    else
      a[i] := intval ( pred (inf (x)), succ (sup (x)) );
  end; {for}

  infl := a;
end;   {operator inflated_by}
```

{- -}

function approximate_inverse (A: rmatrix): rmatrix[1..ubound(A),1..ubound(A)];

{ Computation of an approximate inverse of the (n,n)-matrix A }
{ by Gauss elimination.   }

var
  i, j, k, n : integer;
  factor   : real;
  R, Inv, E : rmatrix[1..ubound(A),1..ubound(A)];
begin
  n := ubound (A);   { dimension of A }

  E := id (E);   { identity matrix }
  R := A;

  { Gaussian elimination step with unit vectors as }
  { right-hand sides. Division by R[i,i]=0 indicates }
  { that the matrix A is probably singular.   }

  for i:= 1 **to** n **do**
    for j:= (i+1) **to** n **do**
    begin
      factor := R[j,i]/R[i,i];
      for k:= i **to** n **do** R[j,k] := #*(R[j,k] − factor*R[i,k]);
      E[j] := E[j] − factor*E[i];
    end;   {for j:= ...}

  { Backward substitution delivers the rows of the inverse of A. }

  for i:= n **downto** 1 **do**
    Inv[i] := #*(E[i] − **for** k:= (i+1) **to** n **sum**(R[i,k]*Inv[k]))/R[i,i];

  approximate_inverse := Inv;
end;   {function approximate_inverse}

{- -}

global procedure linear_system_solver (A : rmatrix; b : rvector;

var x : ivector; **var** ok : boolean);

{ Computation of a verified inclusion vector for the solution of the }
{ linear system of equations. If an inclusion is not achieved after   }
{ a certain number of iteration steps, then the algorithm is stopped, }
{ and the parameter ok is set to false.   }

const
    epsilon   = 0.25; { Constant for the epsilon inflation   }
    max_steps = 10;   { Maximum number of iteration steps }

var
    i   : integer;
    y, z : ivector[1..ubound(A)];
    R   : rmatrix[1..ubound(A),1..ubound(A)];
    C   : imatrix[1..ubound(A),1..ubound(A)];

begin
    R := approximate_inverse (A);

    { R*b is an approximate solution of the linear system and z is an inclusion }
    { of this vector. However, z does not usually include the true solution.   }

    z := R * intval (b);

    { An inclusion of I − R*A is computed with maximum accuracy. }
    { The (n,n) identity matrix is generated by the function call id(A). }

    C := ##(id(A) − R*A);

    x := z;   i := 0;
    repeat
        i := i + 1;

        y := x inflated_by epsilon; { To obtain a true inclusion, the interval }
                                    { vector x is slightly enlarged.   }

        x := z + C*y;   { The new iterate is computed. }

        ok := x **in** y;   { Is x contained in the interior of y? }

    until ok **or** (i = max_steps);
end;   {procedure linear_system_solver}

{- -}

end.   {module lin_solv}

Example

If we use a 10×10 Boothroyd/Dekker matrix (see chapter 5, Exercise 5) to test this program, then the output is:

```
Please enter the dimension n of the linear system: 10
Please enter the matrix A:
   10      45     120     210     252     210     120      45      10       1
   55     330     990    1848    2310    1980    1155     440      99      10
  220    1485    4752    9240   11880   10395    6160    2376     540      55
  715    5148   17160   34320   45045   40040   24024    9360    2145     220
 2002   15015   51480  105105  140140  126126   76440   30030    6930     715
 5005   38610  135135  280280  378378  343980  210210   83160   19305    2002
11440   90090  320320  672672  917280  840840  517440  205920   48048    5005
24310  194480  700128 1485120 2042040 1884960 1166880  466752  109395   11440
48620  393822 1432080 3063060 4241160 3938220 2450448  984555  231660   24310
92378  755820 2771340 5969040 8314020 7759752 4849845 1956240  461890   48620

Please enter the right-hand side b:
1 1 1 1 1 1 1 1 1 1

The matrix A is non-singular. The solution
of the linear system is contained in:

[  9.999999999999998E-001,   1.000000000000001E+000]
[ -1.000000000000001E+000,  -9.999999999999998E-001]
[  9.999999999999998E-001,   1.000000000000001E+000]
[ -1.000000000000001E+000,  -9.999999999999998E-001]
[  9.999999999999998E-001,   1.000000000000001E+000]
[ -1.000000000000001E+000,  -9.999999999999998E-001]
[  9.999999999999998E-001,   1.000000000000001E+000]
[ -1.000000000000001E+000,  -9.999999999999998E-001]
[  9.999999999999998E-001,   1.000000000000001E+000]
[ -1.000000000000001E+000,  -9.999999999999998E-001]
```

Chapter 4

Problem-Solving Routines

Routines for solving common numerical problems have been developed in PASCAL–XSC. They are supplied by means of an additional module library. The methods used compute a highly accurate inclusion of the true solution of the problem and verify the existence and uniqueness of the solution in the given interval. The advantages of these new routines are:

- The solution is computed with high accuracy, even for many ill-conditioned cases.

- The accuracy of the computed solution is always controlled.

- The correctness of the result is automatically verified, i.e. an inclusion set is computed which guarantees the existence and uniqueness of the exact solution within the bounds computed.

- If no solution exists, or if the problem is extremely ill-conditioned, an error message is returned.

PASCAL–XSC routines have been developed for:

- linear systems of equations

    - full systems (*real, complex, interval, cinterval*)
    - matrix inversion (*real, complex, interval, cinterval*)
    - least squares problems (*real, complex, interval, cinterval*)
    - computation of pseudo inverses (*real, complex, interval, cinterval*)
    - band matrices (*real*)
    - sparse matrices (*real*)

- polynomial evaluation

    - in one variable (*real, complex, interval, cinterval*)
    - in several variables (*real*)

- zeros of polynomials (*real, complex, interval, cinterval*)

- eigenvalues and eigenvectors

    - symmetric matrices (*real*)
    - arbitrary matrices (*real, complex, interval, cinterval*)

- initial and boundary value problems of ordinary differential equations

    - linear
    - nonlinear

- evaluation of arithmetic expressions
- nonlinear systems of equations
- numerical quadrature
- integral equations
- automatic differentiation
- optimization problems

For further information about the individual routines and modules, see the documentation enclosed with the PASCAL–XSC numeric library.

In addition to solving the basic problems, these routines can be used for other explorations. They provide answers to rather interesting and important questions, such as

- Determination of the condition of problems by the use of interval input.

- Determination of local exclusion domains (regions in which a solution can be guaranteed *not* to exist).

- Verification of processes such as determination of the minimum rank of a matrix or determination of a sphere or a half plane including all zeros of a complex polynomial. Thus, it is possible, to guarantee stability of technical devices as far as the mathematical model corresponds to reality.

- Parameter control of models. It is easy to determine how sensitively any model data affects a model formula and vice versa. We can compute how accurately the data must be measured in order to guarantee a predefined accuracy for quantities depending on these data.

- In critical cases where there may not be sufficient processing power available to use inclusion methods in real time control problems, it may still be possible to use inclusion methods running in the background to monitor the accuracy and reliability of the foreground processing. A special field of application is the scope of security problems (navigation and control of satellites and aircraft, spacecraft, as well as highly-sensitive, large technical equipment).

Verifying methods are unilateral decision processes which, on the basis of given computing resources (run-time, memory requirements, mantissa length, etc.), detect solvable problems and enclose their solution to a desired accuracy. For example, an enclosure algorithm for solving a system of linear equations may validate the existence of a unique solution and compute an enclosure of the solution. If the algorithm does not succeed, that does not imply that the system is singular. It might be that the same algorithm could solve the problem with more time, memory, or precision. Other algorithms might be used to verify that a matrix is singular.

There is an extensive literature in the area of interval analysis, enclosure methods, and self-validating computation. Neumaier [38] contains an extensive bibliography. The bibliography of our book includes some of the more significant references. Especially noteworthy are the introductory texts [1], [2], [34], [35], and collections of conference papers in [16], [27], [30], [31], [32], [36], [39], [40], [41], [50], and [51].

Chapter 5

Exercises with Solutions

Here are some exercises with which the reader can practice the language PASCAL–XSC by solving various exercises and applying the new language elements to the development of complete programs.

A series of simple exercises is given which use the most important language elements of PASCAL–XSC. This series covers

- introductory exercises

- exercises to go more deeply into the new concepts of PASCAL–XSC (operator concept, functions with arbitrary result type, dynamic arrays, module concept, etc.)

- easy exercises to treat problems of accuracy in arithmetic operations and numerical computations (use of the type *dotprecision*)

- exercises dealing with various arithmetics (interval arithmetic, complex arithmetic, matrix/vector arithmetic, etc.)

- exercises concerning physical and engineering applications of programming languages and numerical methods.

Most of the exercises are taken from a collection that has been developed in connection with lectures on programming languages held at the University of Karlsruhe.

Our proposed solutions, the complete program listings, and some results follow each exercise. The results were produced on an HP 9000 Workstation using an implementation of PASCAL–XSC with 53 bit binary arithmetic.

The results computed with binary arithmetic may differ from the results computed with decimal arithmetic according to the problems of conversion described in chapter 2. In this case, some deviations in the run-time outputs may be possible.

Exercise 1: Test of Representability

Write a PASCAL–XSC program to determine whether a pair of *integer* numbers
z, n ($n \neq 0$) have a quotient z/n which is exactly representable as a *real* number (in
the computer's set of floating-point numbers).

Hint: z/n is exactly representable on the computer if and only if

$$z/< n = z/> n.$$

Your program should loop to read and check an arbitrary number of such pairs. If
the condition is fulfilled, then z, n, and z/n are to be printed. If the condition is not
fulfilled, write a message to that effect. Use $n = 0$ to terminate the loop. After the
termination of the loop, the percentage of the pairs with an exactly representable
quotient is to be computed and reported (rounded to one place past the decimal
point).

Solution:

```
program represent (input, output);

{ Exercise 1: Test of Representability }

var n, z           : integer;
    no_of_exacts,
    no_of_pairs    : integer;
    quotient       : real;

begin
  writeln ('Exercise 1: Test of Representability');
  writeln;
  no_of_exacts:= 0;
  no_of_pairs := 0;
  write ('Enter z and n for quotient test: ');
  read (z, n);
  while n <> 0 do
  begin
    quotient    := z/<n;
    no_of_pairs:= no_of_pairs+1;
    if quotient = z/>n then
      begin
        no_of_exacts := no_of_exacts+1;
        writeln ('Quotient is exactly representable!');
        writeln (z:1,'/',n:1,' = ',quotient);
      end
    else
      writeln ('Quotient is not exactly representable!');
```

```
      writeln;
      write ('Enter z and n for quotient test: ');
      read (z, n);
    end;
  if no_of_pairs <> 0 then
  begin
    writeln;
    writeln (no_of_pairs,  ' data pairs were entered');
    writeln (no_of_exacts, ' quotients are exactly representable');
    writeln ('These are ', no_of_exacts/no_of_pairs*100:5:1:-1,'%');
  end;
end.
```

Exercise 2: Summation of Exponential Series

The function e^x is approximated by a partial sum of its *Taylor* series:

$$S_n = \sum_{i=0}^{n} a_i \quad \text{with} \quad a_i = \frac{x^i}{i!} \quad \text{and} \quad i! = \begin{cases} 1 & \text{for } i = 0 \\ 1 * 2 * \cdots * i & \text{for } i > 0 \end{cases}$$

The partial sum may be computed according to the following algorithm:

$$\text{start:} \quad S_1 = 1; \; a_1 = x$$
$$\text{recursion:} \quad S_i = S_{i-1} + a_{i-1}; \quad a_i = a_{i-1}\frac{x}{i}; \quad i = 2, \ldots, n$$

Write a PASCAL–XSC program with $n = 100$ which reads x and computes S_n using three different rounding controls:

- downwardly directed,

- rounded to the next floating-point number,

- upwardly directed.

The computation of the sum should be terminated before handling the n-th summand if the upwardly directed sum S_i satisfies

$$|a_i| < eps * |S_i| \quad \text{with} \quad eps = 10^{-12}.$$

You should report the correct value of $exp(x)$ and the final values of i and S_i.

Hint: Use negative values (< -50) for x when testing your program since they demonstrate very clearly rounding errors occurring during summation.

Solution:

```
program expo (input, output);

  { Exercise 2: Summation of the exponential series }

const eps = 1e-12;
      n   = 100;

var  Sdown, Snext, Sup : real;
     adown, anext, aup : real;
     i                 : integer;
     x, help           : real;

begin
  writeln ('Exercise 2: Summation of the exponential series');
  writeln;
  write ('Enter an argument x for computing exp(x): ');
  read (x);
```

```
writeln ('Summation of the exponential series :');
writeln ('Step         Summand                   Sum');
adown:= x; anext:= x; aup:= x;
Sdown:= 1; Snext:= 1; Sup:= 1;
i:= 1;
repeat
  i    := i+1;
  Sdown:= Sdown +< adown;
  Snext:= Snext +  anext;
  Sup  := Sup   +> aup;
  anext:= anext *  x /  i;
  if x >= 0 then
    begin
      adown:= adown *< x /< i;
      aup  := aup   *> x /> i;
    end
  else
    begin
      help := adown;
      adown:= aup  *< x /< i;
      aup  := help *> x /> i;
    end;
  writeln (i  :7,' ', adown, ' ', Sdown);
  writeln (' ':7,' ', anext, ' ', Snext);
  writeln (' ':7,' ', aup,  ' ', Sup);
until (i >= n) or (abs(adown) < eps*abs(Sdown));
writeln('Exact value of the function exp(x) : ',exp(x));
end.
```

Exercise 3: Influence of Rounding Errors

Write a PASCAL–XSC program to demonstrate the influence of rounding errors during computation of the expression

$$z = x^4 - 4y^4 - 4y^2$$

for different values of x and y. The program should accept the *real*-values x and y and compute z according to the following methods

1) $z = x \cdot x \cdot x \cdot x - 4 \cdot y \cdot y \cdot y \cdot y - 4 \cdot y \cdot y$ using

    a) the floating point operators * and -

    b) the directed-rounding operators *<, *>, and -< to deliver a lower bound
       for the expression

    c) the directed-rounding operators *<, *>, and -> to deliver an upper bound
       for the expression

2) $z = x^2 \cdot x^2 - 4 \cdot y^2 \cdot y^2 - 4 \cdot y^2$ using the predefined function *sqr* and the
    operators * and -

3) $z = (x^2)^2 - (2 \cdot y^2)^2 - (2 \cdot y)^2$ using *sqr*, *, and -

4) $z = (x^2)^2 - (2 \cdot y)^2 \cdot (y^2 + 1)$ using *sqr*, *, and -

5) $z = \# * (a \cdot a - b \cdot b - c \cdot c)$ with $a = x^2$, $b = 2 \cdot y^2$ and $c = 2 \cdot y$.

The seven computed values should be reported with an accompanying comment for each operation. Test your program by using the values

$$x = 665857.0 \text{ and } y = 470832.0.$$

In this special case, the true value for z is the number 1. For a detailed description of rounding error effects see [52], for example.

Solution:

```
program rounding (input, output);

{ Exercise 3: Influence of Rounding Errors }

var x,y,z: real;
    a,b,c: real;

begin
  writeln('Exercise 3: Influence of Rounding Errors');
  writeln;
  write('x = '); read(x);
  write('y = '); read(y);
```

```
      writeln;
      writeln('Computation of the expression  z = x^4 - 4y^4 - 4y^2');
      writeln;
      z:= x*x*x*x-4*y*y*y*y-4*y*y;
      writeln('Comp.: x*x*x*x-4*y*y*y*y-4*y*y                 = ',z);
      z:= (x*<x)*<(x*<x) -< 4*>(y*>y)*>(y*>y) -< 4*>(y*>y);
      writeln('Comp.: (x*<x)*<(x*<x)-<4*>(y*>y)*>(y*>y)-<4*>(y*>y)= ',z);
      z:= (x*>x)*>(x*>x) -> 4*<(y*<y)*<(y*<y) -> 4*<(y*<y);
      writeln('Comp.: (x*>x)*>(x*>x)->4*<(y*<y)*<(y*<y)->4*<(y*<y)= ',z);
      z:= sqr(x)*sqr(x) - 4*sqr(y)*sqr(y) - 4*sqr(y);
      writeln('Comp.: x^2*x^2-4*y^2*y^2-4*y^2                 = ',z);
      z:= sqr(sqr(x))-sqr(2*sqr(y)) - sqr(2*y);
      writeln('Comp.: (x^2)^2-(2*y^2)^2-(2*y)^2               = ',z);
      z:= sqr(sqr(x))-sqr(2*y) * (sqr(y)+1);
      writeln('Comp.: (x^2)^2-(2*y)^2*(y^2+1)                 = ',z);
      a:=sqr(x);
      b:=2*sqr(y);
      c:=2*y;
      z:=#*(a*a-b*b-c*c);
      writeln('Comp.: #*(x^2*x^2-(2*y^2)*(2*y^2)-(2*y)*(2*y))    = ',z);
   end.
```

Runtime Output:

Influence of rounding errors

```
x = 665857.0
y = 470832.0


Computation of the expression  z = x^4-4y^4-4y^2

Comp.: x*x*x*x-4*y*y*y*y-4*y*y                     =  1.1885568000000E+007
Comp.: (x*<x)*<(x*<x)-<4*>(y*>y)*>(y*>y)-<4*>(y*>y)= -5.5223296000000E+007
Comp.: (x*>x)*>(x*>x)->4*<(y*<y)*<(y*<y)->4*<(y*<y)=  1.1885568000000E+007
Comp.: x^2*x^2-4*y^2*y^2-4*y^2                     =  1.1885568000000E+007
Comp.: (x^2)^2-(2*y^2)^2-(2*y)^2                   =  1.1885568000000E+007
Comp.: (x^2)^2-(2*y)^2*(y^2+1)                     =  0.0000000000000E+000
Comp.: #*(x^2*x^2-(2*y^2)*(2*y^2)-(2*y)*(2*y))     =  1.0000000000000E+000
```

Remark: The #-expression delivers the exact result. This is because all operands
(here a, b, c) are exact since $x = 665875$ and $y = 470832$ are exactly representable.

Exercise 4: Scalar Product

Write a PASCAL–XSC program to compute the value of a scalar product

$$x \cdot y = \sum_{i=1}^{n} x_i \cdot y_i$$

of two real vectors $x, y \in \mathbb{R}^n$. Compare the value computed with maximum accuracy with the value computed in the usual manner.

Write a function *Scalp* producing the scalar product in the usual manner and a function *Max_Acc_Scalp* computing the scalar product by summing up the products $x_i \cdot y_i$ in a variable of the type *dotprecision* and by a single final rounding to a *real* value.

The vectors x and y should be entered in the main program. The values computed via *Scalp* and *Max_Acc_Scalp* should be reported together with comments. Choose $n = 5$ for the declaration of the vector types. Test your program with the vectors

$$x = \begin{pmatrix} 2.718281828E10 \\ -3.141592654E10 \\ 1.414213562E10 \\ 5.772156649E9 \\ 3.010299957E9 \end{pmatrix}, \quad y = \begin{pmatrix} 1.4862497E12 \\ 8.783669879E14 \\ -2.237492E10 \\ 4.773714647E15 \\ 1.85049E5 \end{pmatrix}.$$

The function *Max_Acc_Scalp* of this program simulates the functionality of the operator ∗ for type *rvector* supplied by module MV_ARI.

Solution:

```
program Scalar_Product (input, output);

{ Exercise 4: Scalar Product }

const n = 5;
type  vector = array [1..n] of real;
var   x, y : vector;
      i    : integer;

function Scalp (x, y : vector) : real;
  var s : real;
      i : integer;
  begin
    s:= 0;
    for i:=1 to n do  s:= s + x[i]*y[i];
    Scalp:= s;
  end;
```

```
function Max_Acc_Scalp (x, y : vector) : real;
  var d : dotprecision;
      i : integer;
  begin
    d:= #(0);
    for i:=1 to n do  d:= #(d + x[i]*y[i]);
    Max_Acc_Scalp:= #*(d);
  end;

begin
  writeln('Exercise 4: Scalar Product');
  writeln;
  writeln('Enter 1. vector (with ',n:1,' components):');
  for i:=1 to n do  read(x[i]);
  writeln('Enter 2. vector (with ',n:1,' Components):');
  for i:=1 to n do  read(y[i]);
  writeln;
  writeln('Scalar product in the usual manner : ',Scalp(x,y));
  writeln('Scalar product with dotprecision   : ',Max_Acc_Scalp(x,y));
end.
```

Runtime Output:

```
Exercise 4: Scalar Product

Enter 1. vector (with 5 components):
2.718281828e10
-3.141592654e10
1.414213562e10
5.772156649e9
3.010299957e9
Enter 2. vector (with 5 components):
1.4862497e12
8.783669879e14
-2.237492e10
4.773714647e15
1.85049e5

Scalar product in the usual manner :  4.328386285000000E+009
Scalar product with dotprecision   : -1.006571070000000E+008
```

Exercise 5: Boothroyd/Dekker Matrices

The (integer) elements of a n-dimensional Boothroyd/Dekker matrix (see [53]) $D = (d_{ij})$ are given by

$$d_{ij} = \binom{n+i-1}{i-1} \cdot \binom{n-1}{n-j} \cdot \frac{n}{i+j-1} \quad .$$

Write a PASCAL–XSC program to compute an n-dimensional Boothroyd/Dekker matrix. Use an operator *Choose* for the (integer) computation of the binominal coefficient $\binom{m}{k}$. Write the values of the matrix row by row. The value n (≤ 10) should be entered.

Hint: Compute $\binom{m}{k}$ this way:

$$c_0 := 1; \quad c_i := c_{i-1} * \frac{m-i+1}{i}; \quad i = 1, \dots, k$$

$$\binom{m}{k} := c_k;$$

Use the *integer* division **div**.

Solution:

```
program BDM (input, output);

{ Exercise 5: Boothroyd/Dekker Matrices }

var i, j, n, d : integer;

priority Choose = *;

operator Choose (m, k: integer) ChooseResult : integer;
  var i, c: integer;
  begin
    c:= 1;
    for i:=1 to k do
      c:= (c*(m-i+1)) div i;
    ChooseResult:= c;
  end;

begin
  writeln('Exercise 5: Boothroyd/Dekker Matrices');
  writeln;
  write('Enter the desired dimension of the matrix (<=10): ');
  read (n);
```

```
  writeln;
  for i:=1 to n do
  begin
    for j:=1 to n do
    begin
      d:=(((n+i-1) Choose (i-1))*((n-1) Choose (n-j))*n) div (i+j-1);
      write (d:8);
    end;
    writeln;
  end;
end.
```

Runtime Output:

Exercise 5: Boothroyd/Dekker Matrices

Enter the desired dimension of the matrix (<=10): 8

8	28	56	70	56	28	8	1
36	168	378	504	420	216	63	8
120	630	1512	2100	1800	945	280	36
330	1848	4620	6600	5775	3080	924	120
792	4620	11880	17325	15400	8316	2520	330
1716	10296	27027	40040	36036	19656	6006	792
3432	21021	56056	84084	76440	42042	12936	1716
6435	40040	108108	163800	150150	83160	25740	3432

Exercise 6: Complex Functions

Write a PASCAL–XSC program which simulates some features supplied in module C_ARI to compute the values e^z, $\cos z$, $\sin z$, $\cosh z$, and $\sinh z$ for 20 complex numbers $z = x + iy$ and produces a table of these values.

Use the predefined type *complex* and define

- a monadic operator *I_times* to multiply a complex number by the imaginary unit i,

- a monadic operator $-$ for complex numbers,

- two operators $+$ and $-$ for the addition and subtraction of two complex numbers,

- an operator $*$ for the multiplication of a *real* number by a complex number,

- functions *exp*, *cos*, *sin*, *cosh*, and *sinh* using the predefined functions *sin*, *cos*, and *exp* for *real* quantities, and

- procedures for the input and output.

Hints: $(u, z \in \mathbb{C}; \quad x, y, v, w \in \mathbb{R})$

If $z = x + iy$, then $u = iz$ with $u = v + iw$ is given by $v = -y$ and $w = x$.

Division by i is replaced by a multiplication by $-i$.

real divisions are replaced by multiplications.

If $z = x + iy$, then $e^z = e^x \cos y + ie^x \sin y$ (Euler's formula)

$$\cos z = \frac{e^{iz} + e^{-iz}}{2} \qquad \cosh z = \frac{e^z + e^{-z}}{2}$$

$$\sin z = \frac{e^{iz} - e^{-iz}}{2i} \qquad \sinh z = \frac{e^z - e^{-z}}{2}$$

Solution:

```
program complex_functions (input, output);

{ Exercise 6: Complex Functions }

var
  z : complex;
  c : array [1..20] of complex;
  i : integer;
```

```
priority I_times = ^;
operator I_times (z : complex) multiplied_by_i : complex;
  { Monadic operator for the multiplication of the argument }
  { with the imaginary unit i (with sqr(i) = -1).           }
  begin
    multiplied_by_i.re:= -z.im;
    multiplied_by_i.im:=  z.re;
  end;

operator + (a, b: complex) plus : complex;
  begin
    plus.re:= a.re + b.re;
    plus.im:= a.im + b.im;
  end;

operator - (a, b: complex) minus : complex;
  begin
    minus.re:= a.re - b.re;
    minus.im:= a.im - b.im;
  end;

operator - (a: complex) negate : complex;
  begin
    negate.re:= -a.re;
    negate.im:= -a.im;
  end;

operator * (r: real; z: complex) mulrc : complex;
  begin
    mulrc.re:= r * z.re;
    mulrc.im:= r * z.im
  end;

function exp (z: complex) : complex;
  begin
    exp.re:= exp (z.re) * cos(z.im);
    exp.im:= exp (z.re) * sin(z.im);
  end;

function cos (z: complex) : complex;
  begin
    cos:= 0.5 * (exp (I_times z) + exp (- I_times z))
  end;

function sin (z: complex) : complex;
  begin
    sin:= 0.5 * - I_times (exp (I_times z) - exp (-I_times z))
  end;
```

```pascal
function cosh (z: complex) : complex;
  begin
    cosh:= 0.5 * (exp (z) + exp (-z))
  end;

function sinh (z: complex) : complex;
  begin
    sinh:= 0.5 * (exp (z) - exp (-z));
  end;

procedure write (var f: text; c: complex; s: integer);
  begin
    write (f,'(',c.re:s,',',c.im:s,') ');
  end;

procedure read (var f: text; var c: complex);
  begin
    read (f, c.re, c.im);
  end;

begin
  writeln('Exercise 6: Complex Functions');
  writeln;
  for i:=1 to 20 do
  begin
    write ('Enter ', i:2, '. complex number: ');
    read  (c[i]);
  end;
  writeln;
  writeln(' z ':11,' exp(z)':25,' cos(z) ':25,' sin(z) ':25);
  for i:=1 to 20 do
  begin
    z:= c[i];
    write    (z:8);
    write    (exp(z):8);
    write    (cos(z):8);
    writeln (sin(z):8);
  end;
  writeln;
  writeln(' z ':11,' cosh(z)':25,' sinh(z) ':25);
  for i:=1 to 20 do
  begin
    z:= c[i];
    write    (z:8);
    write    (cosh(z):8);
    writeln (sinh(z):8);
  end;
end.
```

Exercise 7: Surface Area of a Parallelepiped

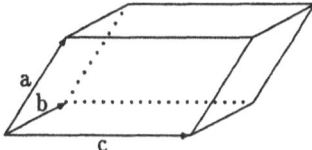

Compute the surface area of a 3-dimensional parallelepiped. We use the following notation

- $a = \begin{pmatrix} a_1 \\ a_2 \\ a_3 \end{pmatrix}$, $b = \begin{pmatrix} b_1 \\ b_2 \\ b_3 \end{pmatrix}$, and $c = \begin{pmatrix} c_1 \\ c_2 \\ c_3 \end{pmatrix}$ are vectors in $\mathbb{R}^3$.

- The scalar product $a \cdot b$ (dot product) of two vectors a and b is defined as

$$a \cdot b = \sum_{i=1}^{3} a_i \cdot b_i = a_1 \cdot b_1 + a_2 \cdot b_2 + a_3 \cdot b_3.$$

- The length L of a vector a is computed as $L(a) := \sqrt{a \cdot a}$.

- The vector product $a \times b$ (cross product) of two vectors yields a vector

$$a \times b := \begin{pmatrix} a_2 \cdot b_3 - a_3 \cdot b_2 \\ a_3 \cdot b_1 - a_1 \cdot b_3 \\ a_1 \cdot b_2 - a_2 \cdot b_1 \end{pmatrix}.$$

- The area of a parallelogram defined by the vectors a and b is $\text{Area}(a, b) := L(a \times b)$.

- The surface area of a parallelepiped defined by the vectors a, b, and c is determined by

$$\text{Surf}(a, b, c) := 2 \cdot (\text{Area}(a, b) + \text{Area}(b, c) + \text{Area}(c, a)).$$

Write a PASCAL–XSC program with these parts:

a) a type *Vector* declared as an array of length 3 whose component type is *real*,

b) an operator * for the scalar product of two vectors,

c) a function *Length* for the length L of a vector

d) an operator *Cross* for the vector product of two vectors,

e) a function *Area* for the area of a parallelogram,

f) a function *Surface* for the surface area of a parallelepiped,

g) a main program which repetitively reads in three vectors from a file with
component type *Vector* and computes the surface of the corresponding parallelepiped. Write the result. The loop should terminate when the end of the
input file is reached.

Hint: The number of vectors in the input file is a multiple of 3.

The vector products b) and d) should be computed with maximum accuracy
via *dotprecision* expressions or accurate expressions.

Solution:

```
program parallelepiped (datafile, input, output);

{ Exercise 7: Surface Area of a Parallelepiped }

type Vector = array [1..3] of real;

operator * (a, b: Vector) scalp : real;
var i : integer;
  begin
    scalp:= #* (for i:=1 to 3 sum (a[i]*b[i]));
  end;

function Length (a: Vector) : real;
  begin
    Length:= sqrt (a*a);
  end;

priority cross = *;

operator cross (a, b: Vector) cprod : Vector;
  begin
    cprod[1] := #*(a[2]*b[3]-a[3]*b[2]);
    cprod[2] := #*(a[3]*b[1]-a[1]*b[3]);
    cprod[3] := #*(a[1]*b[2]-a[2]*b[1]);
  end;

function Area (a, b : VECTOR) : real;
  begin
    Area:= Length (a cross b);
  end;

function Surface (a, b, c : VECTOR) : real;
  begin
    Surface:= 2 * (Area(a,b) + Area(b,c) + Area(c,a));
  end;
```

```
var datafile : file of VECTOR;
    a, b, c  : VECTOR;

begin { main program }
  writeln ('Exercise 7: Surface Area of a Parallelepiped');
  writeln;
  reset (datafile);
  repeat
    read (datafile, a);
    read (datafile, b);
    read (datafile, c);
    writeln ('a : ',a[1],' ',a[2],' ',a[3]);
    writeln ('b : ',b[1],' ',b[2],' ',b[3]);
    writeln ('c : ',c[1],' ',c[2],' ',c[3]);
    write   ('Surface area of the parallelepiped : ');
    writeln (Surface (a,b,c));
  until eof(datafile)
end.
```

Exercise 8: Parallelism and Intersection of Lines

Investigate two lines of the form $a_1 x + b_1 y = c_1$ and $a_2 x + b_2 y = c_2$ in the Euclidean plane for parallelism. In case the two lines are not parallel, find their intersection point S. Write a PASCAL–XSC program including

a) a type declaration *Line* defining a line as an array consisting of 3 components with component type *real*,

b) a type declaration *Point* defining a point as a *record* with the components x and y of type *real*,

c) a real function *Determinant* with the parameters a, b, c, d of type *real* for computing the determinant

$$\det (a, b, c, d) = \begin{vmatrix} a & b \\ c & d \end{vmatrix} = a * d - b * c,$$

using a *dotprecision* expression with rounding away from zero,

d) a logical operator *Parallel_To* delivering the value $TRUE$ for two lines g_1 and g_2 if and only if $\det (a_1, b_1, a_2, b_2) = 0$,

e) an operator ** for the intersection of two variables g_1 given by a_1, b_1, c_1 and g_2 given by a_2, b_2, c_2 of type *Line*. The result is of type *Point* and contains the coordinates of the intersection point $s = (x_s, y_s)$ of the lines g_1 and g_2. Use part c) and the definitions

$$x_s = \frac{\begin{vmatrix} c_1 & b_1 \\ c_2 & b_2 \end{vmatrix}}{\begin{vmatrix} a_1 & b_1 \\ a_2 & b_2 \end{vmatrix}}, \qquad y_s = \frac{\begin{vmatrix} a_1 & c_1 \\ a_2 & c_2 \end{vmatrix}}{\begin{vmatrix} a_1 & b_1 \\ a_2 & b_2 \end{vmatrix}}.$$

f) a main program which uses a loop to read in the parameters of two lines and uses the operator *Parallel_To* to check whether the two lines are parallel. If the lines are parallel, the program should announce, "the lines are parallel". If not, then compute intersection point using part e) and report it. The loop is to be repeated until one of the conditions $a_1 = b_1 = 0$ or $a_2 = b_2 = 0$ hold.

Solution:

```
program parallel (input, output);

{ Exercise 8: Parallelism and Intersection of Lines }

type
  Comp   = (a,b,c);
  Line   = array [Comp] of real;
  Point  = record
              x, y : real;
            end;
var
  g1, g2 : Line;
  s      : Point;

function Determinant (a, b, c, d: real): real;
  var
    dp : dotprecision;
  begin
    dp:= #(a*d - b*c);
    if sign(dp) = 0 then
      Determinant:= 0
    else if sign(dp) < 0 then
      Determinant:= #< (dp)
    else
      Determinant:= #> (dp);
  end;

priority Parallel_To = =;

operator Parallel_To (g1, g2: line) par: boolean;
  begin
    par:= (Determinant(g1[a],g1[b],g2[a],g2[b]) = 0);
  end;

operator ** (g1, g2: line) intersection: point;
  var
    det : real;
  begin
    det:= Determinant (g1[a],g1[b],g2[a],g2[b]);
    intersection.x:= Determinant (g1[c],g1[b],g2[c],g2[b]) / det;
    intersection.y:= Determinant (g1[a],g1[c],g2[a],g2[c]) / det;
  end;

begin {main program}
  writeln ('Exercise 8: Parallelism and Intersection of Lines');
  writeln;
```

```
repeat
  writeln ('Enter the values a1, b1, and c1 for line g1: ');
  read    (g1[a], g1[b], g1[c]);
  writeln;
  writeln ('Enter the values a2, b2, and c2 for line g2: ');
  read    (g2[a], g2[b], g2[c]);
  writeln;
  if g1 Parallel_To g2 then
    writeln ('The two lines are parallel!')
  else
    begin
      s:= g1 ** g2;
      writeln ('The two lines intersect in point ');
      writeln ('(xs,ys) = (',s.x,',',s.y,')');
    end;
  writeln; writeln;
until ((g1[a]=0) and (g1[b]=0)) or ((g2[a]=0) and (g2[b]=0));
end.
```

Exercise 9: Transposed Matrix, Symmetry

An $n \times n$ matrix $A = (a_{ij})$ is called *symmetric* if $a_{ij} = a_{ji}$ for all $i, j \in \{1, \ldots, n\}$. The *transposed* matrix $T = (t_{ij}) = A^T$ of a matrix A is defined by $t_{ij} = a_{ji}$ for all $i, j \in \{1, \ldots, n\}$.

Write a program which

- defines a dynamic type *Matrix* for *integer* matrices,

- defines an operator **=** for two matrices of type *Matrix*,

- declares a monadic operator *Transposed* delivering the transpose of a matrix,

- declares a boolean function *Is_Symmetric* computing the value *TRUE* if and only if its parameter (of the type *Matrix*) is symmetric,

- contains a procedure *Main* declaring two square matrices A and B depending on a parameter n, reading in A and B, ascertaining whether A and B are symmetric or whether $A^T = B$, and reporting the appropriate information,

- reads the value n in the main program and calls the procedure *Main*.

Hint: The function *Is_Symmetric* may be very easily formulated by means of the operators **=** and *Transposed*.

Solution:

```
program Transposed_and_Symmetry (input, output);

{ Exercise 9: Transposed Matrix and Symmetry }

type Matrix = dynamic array [*,*] of real;

operator = (a, b: Matrix) equal: boolean;
  var help : boolean;
      i, j : integer;
  begin
    help:= true;
    for i:= lbound(a,1) to ubound (a,1) do
      for j:= lbound(a,2) to ubound(a,2) do
        if a[i,j] <> b[i,j] then
          help:= false;
    equal:= help;
  end;

priority Transposed = ^;
```

```
operator Transposed (a: matrix) TransposedResult:
        matrix[lbound(a,2)..ubound(a,2),lbound(a,1)..ubound(a,1)];
  var i, j :integer;
  begin
    for i:=lbound(a,2) to ubound(a,2) do
      for j:=lbound(a,1) to ubound(a,1) do
        TransposedResult[i,j]:= a[j,i];
  end;

function Is_Symmetric (a: matrix): boolean;
  begin
    Is_Symmetric:= a = Transposed a;
  end;

procedure read (var f: text; var A: matrix);
  var
    n, i, j: integer;
  begin
    n:= ubound (A);
    for i:=1 to n do
    begin
      write (i:3,'. row: ');
      for j:=1 to n do
        read (A[i,j]);
    end;
  end;

procedure write (var f: text; var A: matrix);
  var
    n, i, j: integer;
  begin
    n:= ubound (A);
    for i:=1 to n do
    begin
      for j:=1 to n do
        write (A[i,j]:5:1);
      writeln;
    end;
  end;

procedure Main (n: integer);
  var i, j     : integer;
      A, B, At : matrix[1..n,1..n];
  begin
    writeln('Enter the elements of matrix A:');
    read (A);
    writeln('Enter the elements of matrix B:');
    read (B);
```

```
      writeln ('Transposed of the matrix A:');
      At:= Transposed A;
      writeln (At);
      if Is_Symmetric (A) then
        writeln('A is symmetric ');
      if Is_Symmetric (B) then
        writeln('B is symmetric ');
      if At = B then
        writeln('Transposed (A) is equal to B ');
  end;

var n : integer;

begin
  writeln('Exercise 9: Transposed Matrix and Symmetry');
  writeln;
  write('Enter the dimension of the matrices: ');
  read (n);
  Main (n);
end.
```

Exercise 10: Rail Route Map

For a German railroad line, a rail route map is to be prepared. Beginning at the starting point *Place_0*, a destination station *Place_9* has to be reached by visiting 8 intermediate stations *Place_1*,...,*Place_8* with a two-minute stop at every station.

Write a program which reads in the time of departure from *Place_0* and the distances between the stations *Place_i* and *Place_i+1* for $i = 0,...,8$. Based upon an average speed of 115 km/h, compute the times of arrival and departure at the stations. Print a rail route map which uses a 24-hour clock!

For this purpose, define

a) a type *Time* as a *record* with the components *Hour* and *Minute*,

b) a function *RouteTime* computing the railroad time required by the rail route section,

c) an operator for the addition of the railroad time and duration of stay to the current time,

d) a main program reading in the necessary data, computing railroad time and duration of stay, and producing a table consisting of columns for place, times of arrival and departure, and distance to the next station.

Hint: In order to compute railroad times within the function *RouteTime* and to implement the operator, it is rather useful to compute in seconds and to convert them to whole minutes.

Solution:

```
program map (input, output);

{ Exercise 10: Rail Route Map }

type
  Time = record
            Hour   : 0..23;
            Minute: 0..59;
         end;

var
  i             : integer;
  route         : array [1..9] of real;
  curr_time,
  departure,
  arrival,
  stop_time    : Time;
```

```
function RouteTime (place_i: integer) : Time;
  var
    hours : real;
    help  : Time;

  begin
    hours      := route[place_i]/115;
    help.Hour  := trunc(hours);
    hours      := hours - help.Hour;
    help.Minute:= trunc(hours*60);
    RouteTime  := help;
  end;

operator + (a, b: Time) sm: Time;
  var
    help : 0..119;

  begin
    help      := a.Minute + b.Minute;
    sm.Minute:= help mod 60;
    help      := a.Hour + b.Hour + help div 60;
    sm.Hour  := help mod 24
  end;

begin
  writeln ('Exercise 10: Rail Route Map');
  writeln;
  stop_time.Hour  := 0;
  stop_time.Minute:= 2;
  write('Enter the departure time (hh mm): ');
  read (departure.Hour, departure.Minute);
  for i:=1 to 9 do
  begin
    write ('Enter the distance between station ',
           i-1:1, ' and ', i:1, ' (in km) : ');
    read (route[i]);
  end;
  writeln;
  writeln('Station     Arrival   Departure    Distance to next station');
  writeln('===========================================================');
  curr_time := departure;
  writeln('PLACE_', '0         --:--       ',
          curr_time.Hour:2,   ':',   curr_time.Minute:2,
          '               ', route[1]:10:2, ' km');
```

```
for i:=1 to 8 do
begin
  arrival    := curr_time + RouteTime(i);
  curr_time := arrival + stop_time;
  writeln('PLACE_', i:1, '          ',
          arrival.Hour:2,    ':',   arrival.Minute:2, '          ',
          curr_time.Hour:2,  ':',   curr_time.Minute:2,
          '            ', route[i+1]:10:2,' km');
end;
arrival := curr_time + RouteTime(9);
writeln('PLACE_','9         ',
        arrival.Hour:2, ':',   arrival.Minute:2, '          ',
                        '--:--',
        '                    --.-- km');
end.
```

Runtime Output:

Station	Arrival	Departure	Distance to next Station
PLACE_0	--:--	10:00	12.00 km
PLACE_1	10:06	10:08	23.00 km
PLACE_2	10:20	10:22	34.00 km
PLACE_3	10:40	10:42	45.00 km
PLACE_4	11:05	11:07	56.00 km
PLACE_5	11:37	11:39	67.00 km
PLACE_6	12:14	12:16	78.00 km
PLACE_7	12:56	12:58	89.00 km
PLACE_8	13:45	13:47	91.00 km
PLACE_9	14:34	--:--	--.-- --

Exercise 11: Inventory Lists

Write a PASCAL–XSC program summing up several individual inventory lists of branches of a chain of department stores in one total list. Use

- a linear linked list (*pointer*) with elements consisting of the components *Ident* (*string* with a maximum of 20 characters) and *Number* (0..*maxint*),

- a procedure for entering of an inventory list,

- an operator + for summing up two lists into one single list by addition of the two *Number* components with the same label or by inserting new list elements,

- a procedure to print the complete inventory list in tabular form.

In the main program, first the number n of the individual lists and then the n lists themselves should be entered. Finally, use of the operator + to sum the individual lists into one single list. Report the inventory in a tabular form.

Solution:

```
program lists (input, output);

{ Exercise 11: Investory Lists }

type
  goods_pointer = ^goods;
  goods         = record
                    ident  : string[20];
                    number : 0..maxint;
                    next   : goods_pointer;
                  end;

procedure list_input (var list: goods_pointer);
  var
    h : goods_pointer;
  begin
    list:= nil;
    repeat
      new (h);
      write ('ident: ');
      readln;
      read (h^.ident);
      write ('number: ');
      read (h^.number);
      h^.next:= list;
      list:= h;
    until list^.number < 0;
```

```
      list:= list^.next;
    end;

operator + (list1, list2: goods_pointer) total_list: goods_pointer;
  var
    total, h1, h2 : goods_pointer;
    flag          : boolean;
  begin
    if (list1 = nil) then
      total := list2
    else
    begin
      total := list1;
      h2    := list2;
      while h2 <> nil do
      begin
        h1 := total;
        repeat
          flag:= h1 <> nil;
          if flag then
            flag:= h1^.ident <> h2^.ident;
          if flag then
            h1:= h1^.next;
        until not flag;
        if h1 <> nil then
        begin
          h1^.number := h1^.number + h2^.number;
          list2      := list2^.next;
        end
        else
        begin
          list2     := list2^.next;
          h2^.next := total;
          total    := h2;
        end;
        h2 := list2;
      end;
    end;
    total_list := total;
  end;

procedure list_output (list: goods_pointer);
  var
    h : goods_pointer;
  begin
    h := list;
    writeln('ident                    number ');
    repeat
```

```pascal
        writeln (h^.ident:24, h^.number);
          h := h^.next;
        until h = nil;
      end;

var
  n, i        : integer;
  list, total : goods_pointer;

begin  {main program}
  writeln('Exercise 11: Investory Lists');
  writeln;
  total := nil;
  write('How many individual lists do you want to enter? ');
  read (n);
  writeln;
  for i:=1 to n do
  begin
    writeln (i:3,'. inventory list:');
    list_input (list);
    writeln;
    list_output (list);
    writeln;
    total:= total + list;
  end;
  writeln;
  writeln(' *** Here is the total list *** ');
  list_output (total);
end.
```

Exercise 12: Complex Numbers and Polar Representation

A complex number $z = a + ib = (a, b)$ with $a, b \in \mathbb{R}$ can be represented in polar coordinates as $z = re^{i\varphi} = (r, \varphi)$ with $r, \varphi \in \mathbb{R}$, $0 \le \varphi < 2\pi$.

Write a PASCAL–XSC program working with this representation. Proceed as follows:

a) Define an appropriate record-type polar_complex with the components r and phi for the representation of complex numbers in polar coordinates.

b) Write a function pi yielding the value π (hint: $\pi = 4 \arctan(1)$).

c) Overload the assignment operator := to enable the assignment of a complex number $z = a + ib$ of type complex to a variable of type polar_complex with components r and phi. The type conversion has to be done according to the formulas

$$r = \sqrt{a^2 + b^2}$$

$$\varphi = \begin{cases} \pi/2 & \text{for } a = 0 \text{ and } b \ge 0 \\ 3/2 * \pi & \text{for } a = 0 \text{ and } b < 0 \\ \arctan(b/a) & \text{for } a > 0 \text{ and } b \ge 0 \\ 2 * \pi + \arctan(b/a) & \text{for } a > 0 \text{ and } b < 0 \\ \pi + \arctan(b/a) & \text{for } a < 0. \end{cases}$$

To compute r, use a #-expression in order to increase accuracy (as far as possible).

d) Define an operator * to compute the product $w = (r, \varphi) = u * v$ of two complex numbers $u = (r_1, \varphi_1)$ and $v = (r_2, \varphi_2)$ of type polar_complex in polar representation by

$$r = r_1 * r_2, \quad \text{and}$$

$$\varphi = \begin{cases} \varphi_1 + \varphi_2 & \text{for } \varphi_1 + \varphi_2 < 2\pi \\ \varphi_1 + \varphi_2 - 2 * \pi & \text{otherwise.} \end{cases}$$

e) Define an operator / to compute the quotient $w = (r, \varphi) = u/v$ of two complex numbers $u = (r_1, \varphi_1)$ and $v = (r_2, \varphi_2)$ of type polar_complex in polar representation by

$$r = r_1/r_2, \quad \text{and}$$

$$\varphi = \begin{cases} \varphi_1 - \varphi_2 & \text{for } \varphi_1 - \varphi_2 \ge 0 \\ \varphi_1 - \varphi_2 + 2 * \pi & \text{otherwise.} \end{cases}$$

f) Write a main program which

    1. reads in two complex numbers u and v of type *complex*,

    2. produces the corresponding values pu and pv using the overloaded assignment operator,

    3. computes the values $w = u * v / u / v$ and $pw = pu * pv / pu / pv$,

    4. reports the radius r and the angle φ of pw, and

    5. reports $pw2 = pcompl(w)$ as a comparison.

Hint: Use the predefined type *complex* and link the module C_ARI providing the operators and input/output procedures which are necessary for this type.

Solution:

```
program polar (input, output);

{ Exercise 12: Complex Numbers and Polar Representation }

use c_ari;

type
  polar_complex = record
                    r, phi : real;
                  end;
var
  u, v, w            : complex;
  pu, pv, pw, pw2 : polar_complex;

function pi : real;
  begin
    pi:= 4 * arctan (1);
  end;

operator := (var pz: polar_complex; z: complex);
  var
    a, b, ph : real;
  begin
    a    := z.re;
    b    := z.im;
    pz.r := sqrt ( #*(a * a + b * b) );
    if (a = 0) and (b >= 0) then
      ph := pi/2
    else if (a = 0) and (b < 0) then
      ph := 3 / 2 * pi
    else if (a > 0) and (b >= 0) then
      ph := arctan (b/a)
```

```
    else if (a > 0) and (b < 0) then
      ph := 2 * pi + arctan (b/a)
    else
      ph := pi + arctan (b/a);
    pz.phi := ph;
  end;

operator * (u, v : polar_complex) resmul : polar_complex;
  var
    ph : real;
  begin
    resmul.r := u.r * v.r;
    ph        := u.phi + v.phi;
    if ph < 2 * pi then
      resmul.phi := ph
    else
      resmul.phi := ph - 2 * pi;
  end;

operator / (u, v : polar_complex) resdiv : polar_complex;
  var
    ph : real;
  begin
    resdiv.r := u.r / v.r;
    ph        := u.phi - v.phi;
    if ph >= 0 then
      resdiv.phi := ph
    else
      resdiv.phi := ph + 2 * pi;
  end;

begin  {main program}
  writeln ('Exercise 12: Complex Numbers and Polar Representation');
  writeln;
  write ('Enter complex number u: ');
  read (u);
  write ('Enter complex number v: ');
  read (v);
  pu   := u;
  pv   := v;
  w    := u * v / u / v;
  pw2 := w;
  pw   := pu * pv / pu / pv;
  writeln ('Radius of pw : ',pw.r);
  writeln ('Angle  of pw : ',pw.phi);
  writeln ('Radius of pw2: ',pw2.r);
  writeln ('Angle  of pw2: ',pw2.phi);
end.
```

Exercise 13: Complex Division

The quotient of two complex numbers $z_1 = x_1 + iy_1$ and $z_2 = x_2 + iy_2$ may be computed as

$$\frac{z_1}{z_2} = \frac{z_1 \overline{z_2}}{z_2 \overline{z_2}} = \frac{(x_1 + iy_1)(x_2 - iy_2)}{x_2^2 + y_2^2} = \frac{x_1 x_2 + y_1 y_2}{x_2^2 + y_2^2} + i\frac{y_1 x_2 - x_1 y_2}{x_2^2 + y_2^2}$$

a) Write a PASCAL–XSC program which includes the declaration of an operator *Cdiv* realizing this complex division for two complex numbers of the type *complex* by application of the operators $+, -, *, /$ for *real* numbers. In the main program, two numbers of type *complex* should be read in, divided, and the result should be printed.

b) Extend your program in such a manner that you link the module *C_ARI*. Call your operator *Cdiv* and then the operator / predefined in *C_ARI*. Compare the output of the two values.

c) Test your program by using the values $z_1 = x_1 + iy_1$ and $z_2 = x_2 + iy_2$ with

$$x_1 = 1254027132096, \quad y_1 = 886731088897$$
$$x_2 = 886731088897, \quad y_2 = 627013566048 \quad .$$

You will notice a clear difference in the imaginary parts of the results.

d) Design another operator *NewCdiv* yielding better results than *Cdiv* by the use of accurate expressions. Compare the three operators in a test run once more.

Solution:

```
program ComplDiv (input, output);

{ Exercise 13: Complex Division }

use c_ari;

var z1, z2 : complex;

priority Cdiv = *;

operator Cdiv (z1, z2 : complex) result : complex;
   var denom : real;
   begin
       denom    := sqr(z2.re) + sqr(z2.im);
       result.re:= (z1.re*z2.re + z1.im*z2.im)/denom;
       result.im:= (z2.re*z1.im - z1.re*z2.im)/denom;
   end;
```

```
priority NewCdiv = *;

operator NewCdiv (z1, z2 : complex) NewResult : complex;
   var denom : real;
   begin
     denom        := #*(z2.re*z2.re + z2.im*z2.im);
     NewResult.re:= #*(z1.re*z2.re + z1.im*z2.im)/denom;
     NewResult.im:= #*(z2.re*z1.im - z1.re*z2.im)/denom;
   end;

begin
  writeln ('Exercise 13: Complex Division');
  writeln;
  write ('Numerator   z1 = '); read (z1);
  write ('Denominator z2 = '); read (z2);
  writeln;
  write ('z1   Cdiv z2 = '); writeln (z1   cdiv z2);
  write ('z1 NewCdiv z2 = '); writeln (z1 NewCdiv z2);
  write ('z1    /    z2 = '); writeln (z1    /    z2);
end.
```

Runtime Output:

```
Exercise 13: Complex Division

Numerator   z1 = ( 1254027132096, 886731088897 )
Denominator z2 = (  886731088897, 627013566048 )


z1   Cdiv z2 = (  1.414213562373095E+000,  0.000000000000000E+000 )
z1 NewCdiv z2 = (  1.414213562373095E+000,  8.478614131951457E-025 )
z1    /    z2 = (  1.414213562373095E+000,  8.478614131951457E-025 )
```

Exercise 14: Electric Circuit

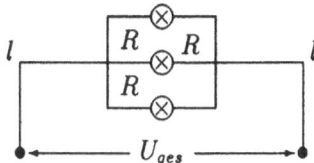

One to three light bulbs with a resistance of $R = 240\Omega$ are connected to a voltage $U = 220V$ via a wire of the length $l = 100m$ with the diameter $d = 1.5mm$ and a specific resistance of $\rho = 0.02857\Omega mm^2/m$. All data are given with an error of 0.5%.

Write a PASCAL–XSC program using the module I_ARI and the type *interval*. For three cases (1, 2, 3 bulbs connected), compute intervals for the range of the values R_{tot} (total resistance of the circuit), I_{tot} (total current), U_w (portion of the voltage at the wire) and U_{bu} (bulb voltage). Use the formulas

$$
\begin{aligned}
R_{tot} &= R_l + R/n \\
I_{tot} &= U/R_{tot} \\
U_w &= R_w \cdot I_{tot} \\
U_{bu} &= U - U_w \quad .
\end{aligned}
$$

Proceed as follows:

1) Read the values π, R, l, d, ρ, and U. Use the error of 0.5% to compute the intervals R, L, D, Rho, and U which enclose the given values. Compute PI enclosing π. Print the values of the intervals.

2) Compute and print the inclusion Rw of the wire resistance

$$ R_w = (8 \cdot \rho \cdot l)/(\pi \cdot d^2). $$

3) Compute and report the intervals Rtot, Itot, Uw, and Ubu for $n = 1, 2, 3$.

4) Finally, print the interval Us representing the total range of Ubu for the different number of bulbs.

Hint: An interval inclusion PI for π is computed by

$$ PI = 4 \cdot \arctan\ ([1,1]). $$

Use the predefined type *interval* and the module I_ARI containing the necessary interval operators.

Solution:

```
program circuit (input, output);

{ Exercise 14: Electric Circuit }

use i_ari;

var
  error,
  U, L, D, Rho, R, Us, pi,
  Rw, Rtot, Itot, Uw, Ubu : interval;
  n                       : integer;

procedure write (var f: text; int: interval;  long: boolean);
  { Overloading of write to allow output of intervals with }
  { a long mantissa. The default output of intervals is    }
  { made according to the width of the interval.           }
  begin
    if long then
      write (f,'[',int.inf:20:0:-1 ,',',int.sup:20:0:+1 ,']')
      { Output with more digits }
    else
      write (f,int);
      { Default output predefined in I_ARI }
  end;

begin
  writeln ('Exercise 14: Electric Circuit');
  writeln;
  write ('U   = '); read (u);
  write ('L   = '); read (l);
  write ('D   = '); read (d);
  write ('Rho = '); read (rho);
  write ('R   = '); read (r);
  pi := 4 * arctan (intval(1));
  error := intval ( (<0.995) , (>1.005) );
  r   := r * error;
  l   := l * error;
  d   := d * error;
  Rho:= Rho * error;
  u   := u * error;
  writeln;
  writeln ('Intervals:');
  writeln ('PI   = ',pi : true);
  writeln ('R    = ',R  : true);
  writeln ('L    = ',L  : true);
  writeln ('D    = ',D  : true);
```

```
  writeln ('Rho  = ',Rho: true);
  writeln ('U    = ',U  : true);
  writeln;
  Rw:= (8*Rho*L)/(pi*sqr(D));
  writeln ('Inclusion of the wire resistance:');
  writeln ('Rw   = ',Rw : true);
  for n:= 1 to 3 do
  begin
    writeln;
    Rtot:= Rw + r/n;
    Itot:= U / Rtot;
    Uw  := Rw * Itot;
    Ubu := U - Uw;
    if n = 1 then
      Us:= Ubu
    else
      Us:= Us +* Ubu;
    write('With ',n:1,' bulb');
    if n <> 1 then write ('s');
    writeln(', the following inclusions are computed:');
    writeln('- total resistance:  Rtot = ',Rtot : true);
    writeln('- total current:     Itot = ',Itot : true);
    writeln('- wire voltage:      Uw   = ',Uw   : true);
    writeln('- bulb voltage:      Ubu  = ',Ubu  : true);
    write('press return'); readln; writeln;
  end;
  writeln ('The bulb voltage Ubu has the total range interval');
  writeln (Us : true);
end.
```

Runtime Output:

Exercise 14: Electric Circuit

```
U   = 220
l   = 100
d   = 1.5
rho = 0.02857
R   = 240
```

Intervals:

```
PI  = [ 3.141592653589E+000, 3.141592653590E+000]
R   = [ 2.387999999999E+002, 2.412000000001E+002]
L   = [ 9.949999999999E+001, 1.005000000001E+002]
D   = [ 1.492499999999E+000, 1.507500000001E+000]
Rho = [ 2.842714999999E-002, 2.871285000001E-002]
```

```
U    = [ 2.188999999999E+002, 2.211000000001E+002]
```

Inclusion of the wire resistance:
```
Rw   = [ 3.169435182649E+000, 3.298783385817E+000]
```

With 1 bulb, the following inclusions are computed:
- total resistance: Rtot = [2.419694351826E+002, 2.444987833859E+002]
- total current: Itot = [8.953009784698E-001, 9.137517713058E-001]
- wire voltage: Uw   = [2.837598420222E+000, 3.014269161944E+000]
- bulb voltage: Ubu = [2.158857308380E+002, 2.182624015798E+002]
press return

With 2 bulbs, the following inclusions are computed:
- total resistance: Rtot = [1.225694351826E+002, 1.238987833859E+002]
- total current: Itot = [1.766764725351E+000, 1.803875490416E+000]
- wire voltage: Uw   = [5.599646279992E+000, 5.950594497863E+000]
- bulb voltage: Ubu = [2.129494055021E+002, 2.155003537201E+002]
press return

With 3 bulbs, the following inclusions are computed:
- total resistance: Rtot = [8.276943518264E+001, 8.369878338582E+001]
- total current: Itot = [2.615330726982E+000, 2.671275930688E+000]
- wire voltage: Uw   = [8.289121220362E+000, 8.811960659084E+000]
- bulb voltage: Ubu = [2.100880393409E+002, 2.128108787797E+002]
press return
```

The bulb voltage Ubu has the total range interval
```
[2.100880393409E+002, 2.182624015798E+002]
```
```

Exercise 15: Alternating Current Measuring Bridge

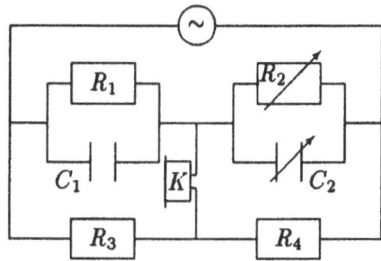

The capacity of the unknown capacitance C_1 and the resistance of the unknown resistor R_1 may be determined by the circuit diagrammed above. This is done by varying the capacity C_2 and the resistance R_2 until the sound in the loudspeaker K reaches a minimum or vanishes. Then, the capacitances and the resistances satisfy

$$C_1 = R_4 \cdot C_2 / R_3$$
$$R_1 = R_3 \cdot R_2 / R_4 \quad .$$

According to the data supplied by the manufacturer, the values of R_3 and R_4 satisfy

$$9.9\Omega \leq R_3 \leq 10.1\Omega$$
$$6.8\Omega \leq R_4 \leq 6.9\Omega \quad .$$

Due to uncertainties of perception, we obtain the estimates for C_2 and R_2

$$40.2\text{mF} \leq C_2 \leq 41.5\text{mF}$$
$$18.3\Omega \leq R_2 \leq 19.8\Omega \quad .$$

Write a PASCAL–XSC program that

- reads in the boundary values of R_3, R_4, C_2, and R_2,

- computes and prints the interval enclosures of C_1 and R_1, and

- repeats the computation of enclosures of C_1 and R_1 assuming that the amount of the error for C_2 and R_2 is 10% higher.

Hint: The 10% increase of the errors for C_2 and R_2 should be handled by enlarging the radius of the corresponding intervals by 10%.

Solution:

```
program measure_bridge (input, output);

{ Exercise 15: Alternating Current Measuring Bridge }

use i_ari;

var
  c1, c2, r1, r2, r3, r4 : interval;
  d : real;

procedure write (var f: text; int: interval;  long: boolean);
  { Overloading of write to allow output of intervals with }
  { a long mantissa. The default output of intervals is    }
  { made according to the width of the interval.           }
  begin
    if long then
      write (f,'[',int.inf:20:0:-1 ,',',int.sup:20:0:+1 ,']')
      { Output with more digits }
    else
      write (f,int);
      { Default output predefined in I_ARI }
  end;

begin
  writeln ('Exercise 15: Alternating Current Measuring Bridge');
  writeln;
  write('lower bound of R3: '); read(r3.inf:-1);
  write('upper bound of R3: '); read(r3.sup:+1);
  write('lower bound of R4: '); read(r4.inf:-1);
  write('upper bound of R4: '); read(r4.sup:+1);
  write('lower bound of C2: '); read(c2.inf:-1);
  write('upper bound of C2: '); read(c2.sup:+1);
  write('lower bound of R2: '); read(r2.inf:-1);
  write('upper bound of R2: '); read(r2.sup:+1);
  { Computing C1 and R1 }
  c1:= r4 * c2 / r3;
  r1:= r3 * r2 / r4;
  writeln;
  writeln('C1 = ', c1 : true);
  writeln('R1 = ', r1 : true);
  { Compute: "10% of the radius", that is "diameter / 20" }
  d := diam(c2) /> 20;
  { Enlarge the interval radius to that amount }
  c2:= intval ( c2.inf -< d , c2.sup +> d );
  { Compute: "10% of the radius", that is "diameter / 20" }
  d := diam(r2) /> 20;
```

```
{ Enlarge the interval radius to that amount }
r2:= intval ( r2.inf -< d , r2.sup +> d );
{ Computing C1 and R1 }
c1:= r4 * c2 / r3;
r1:= r3 * r2 / r4;
writeln;
writeln('Results, with the error for C2 and R2 10% higher:');
writeln;
writeln('C1 = ', c1 : true);
writeln('R1 = ', r1 : true);
end.
```

Runtime Output:

Exercise 15: Alternating Current Measuring Bridge

```
lower bound of R3: 9.9
upper bound of R3: 10.1
lower bound of R4: 6.8
upper bound of R4: 6.9
lower bound of C2: 40.2
upper bound of C2: 41.5
lower bound of R2: 18.3
upper bound of R2: 19.8


C1 = [ 2.706534653465E+001, 2.892424242425E+001]
R1 = [ 2.625652173913E+001, 2.940882352942E+001]

Results, with the error for C2 and R2 10% higher:

C1 = [ 2.702158415841E+001, 2.896954545455E+001]
R1 = [ 2.614891304347E+001, 2.952022058824E+001]
```

Remark: This exercise illustrates how interval arithmetic may be easily applied to error computations in engineering.

Exercise 16: Optical Lens

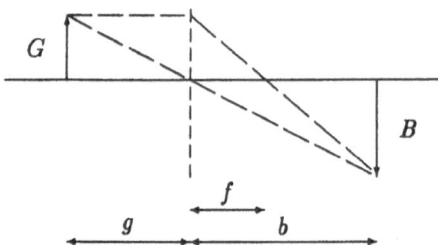

With a lens having a focal length $f = (20 \pm 1)$ cm, an image distance $b = (25 \pm 1)$ cm was measured for the image B of the object G. The lens equation to determine the object distance g of a thin lens is given by

$$\frac{1}{f} = \frac{1}{b} + \frac{1}{g} \ .$$

Hence, g satisfies the equation

$$g = \frac{1}{\frac{1}{f} - \frac{1}{b}}.$$

Usually, the value $g = g_0 \pm \Delta g$ is computed with the approximation g_0 and the error term Δg. This is done by

$$g_0 = \frac{1}{\frac{1}{f_0} - \frac{1}{b_0}}$$

and by the linearization

$$\Delta g = \frac{\Delta f}{(1 - \frac{f_0}{b_0})^2} + \frac{\Delta b}{(\frac{b_0}{f_0} - 1)^2}.$$

Let $f_0 = 20$cm, $b_0 = 25$cm, and $\Delta f = \Delta b = 1$cm. Write a PASCAL–XSC program that

- reads the values for f_0, b_0, Δf, and Δb,

- calculates the interval $g = g_0 \pm \Delta g$ by the method described above,

- calculates the interval g from the intervals f and b applying interval arithmetic according to

$$g = \frac{1}{\frac{1}{f} - \frac{1}{b}},$$

  and

- prints f, b, and the two different values of g along with appropriate comments.

Does the usual method deliver a correct result? Compare it with the enclosure for
g computed with interval arithmetic.

Solution:

```
program opt_lens (input, output);

{ Exercise 16: Optical Lens }

use i_ari;

var
  g0, dg, f0, df, b0, db : real;
  g, f, b                : interval;

procedure write (var f: text; int: interval;  long: boolean);
  { Overloading of write to allow output of intervals with }
  { a long mantissa. The default output of intervals is    }
  { made according to the width of the interval.           }
  begin
    if long then
      write (f,'[',int.inf:20:0:-1 ,',',int.sup:20:0:+1 ,']')
      { Output with more digits }
    else
      write (f,int);
      { Default output predefined in I_ARI }
  end;

begin
  writeln ('Exercise 16: Optical Lens');
  writeln;
  write('f0 = '); read(f0);
  write('df = '); read(df);
  write('b0 = '); read(b0);
  write('db = '); read(db);
  writeln;
  f:= intval (f0 - df , f0 + df);
  b:= intval (b0 - db , b0 + db);
  writeln ('f = ', f : true);
  writeln ('b = ', b : true);
  writeln;
  g0 := 1 / (1/f0 - 1/b0);
  dg := df / sqr(1 - f0/b0) + db / sqr(b0/f0 - 1);
  g  := intval (g0 - dg , g0 + dg);
  writeln ('g = g0 +/- dg       = ', g : true);
  g:= 1/(1/f - 1/b);
  writeln ('g = 1 / (1/f - 1/b) = ', g : true);
end.
```

Runtime Output:

Exercise 16: Optical Lens

```
f0 = 20
df = 1
b0 = 25
db = 1


f = [ 1.900000000000E+001, 2.100000000000E+001]
b = [ 2.400000000000E+001, 2.600000000000E+001]

g = g0 +/- dg       = [ 5.899999999999E+001, 1.410000000000E+002]
g = 1 / (1/f - 1/b) = [ 7.057142857142E+001, 1.680000000001E+002]
```

Remark: The method normally used for error evalution calculates an interval which
is incorrect.

Hint: Exercises 14, 15, and 16 were inspired by the contribution *Technical Calculations by Means of Interval Mathematics*, by P. Thieler [48].

Exercise 17: Interval Evaluation of a Polynomial

Write a PASCAL–XSC program which uses the module I_ARI to evaluate the polynomial

$$p(X) = 1 + 3X - 10X^2$$

using interval arithmetic. Compare the results for the following representations (with X of type interval):

1) $(1 - 2 * X) * (1 + 5 * X)$

2) $1 + 3 * X - 10 * \mathrm{sqr}(X)$

3) $1 + X * (3 - 10 * X)$ (Horner sheme)

4) $1 + 3 * m(X) - 10 * \mathrm{sqr}(m(X)) + (3 - 20 * X)(X - m(X))$
   (mean value form (see [43]), with $m(X)$ as the midpoint of X)

As examples for X, choose both narrow intervals (about one unit in the 14th decimal place) and intervals with other diameters. Intervals around the zeros ($x = 0.5$ and $x = -0.2$) and around the extreme value ($x = 0.15$) should also be considered.

Hint: For the evaluation of the midpoint of an interval, implement a function using an #-expression to obtain maximum accuracy.

Solution:

```
program int_poly (input, output);

{ Exercise 17: Interval Evaluation of a Polynomial }

use i_ari;

var x : interval;

function midpoint (x : interval) : real;

  begin
    midpoint:= #* (0.5 * x.inf + 0.5 * x.sup);
  end;

begin
  writeln ('Exercise 17: Interval Evaluation of a Polynomial');
  writeln;
  repeat
    write ('Enter X : '); read(x);
    writeln;
    writeln ('Method 1: p(X) = ', (1-2*x)*(1+5*x));
```

```
    writeln ('Method 2: p(X) = ', 1+3*x-10*sqr(x));
    writeln ('Method 3: p(X) = ', 1+x*(3-10*x));
    writeln ('Method 4: p(X) = ', 1+3*midpoint(x)
                             -10*sqr(midpoint(x))
                             +(3-20*x)*(x-midpoint(x)));
    writeln; writeln;
  until x = 0;
end.
```

Runtime Output:

Exercise 17: Interval Evaluation of a Polynomial

Enter X : [0.5,0.5]

```
Method 1: p(X) = [  0.000000000000000E+000,  0.000000000000000E+000]
Method 2: p(X) = [  0.000000000000000E+000,  0.000000000000000E+000]
Method 3: p(X) = [  0.000000000000000E+000,  0.000000000000000E+000]
Method 4: p(X) = [  0.000000000000000E+000,  0.000000000000000E+000]
```


Enter X : [0.4999999999,0.5]

```
Method 1: p(X) = [                 0.0E+000,                  7.1E-010]
Method 2: p(X) = [                -3.1E-010,                  1.1E-009]
Method 3: p(X) = [                 0.0E+000,                  7.1E-010]
Method 4: p(X) = [                 0.0E+000,                  7.1E-010]
```


Enter X : [-0.2000000000001,-0.1999999999999]

```
Method 1: p(X) = [                -7.1E-013,                  7.1E-013]
Method 2: p(X) = [                -7.1E-013,                  7.1E-013]
Method 3: p(X) = [                -7.1E-013,                  7.1E-013]
Method 4: p(X) = [                -7.1E-013,                  7.1E-013]
```


Enter X : [0.1499999999999,0.1500000000001]

```
Method 1: p(X) = [    1.224999999999E+000,    1.225000000001E+000]
Method 2: p(X) = [    1.224999999999E+000,    1.225000000001E+000]
Method 3: p(X) = [    1.224999999999E+000,    1.225000000001E+000]
Method 4: p(X) = [  1.224999999999999E+000,  1.225000000000001E+000]
```


Enter X : [0.1,0.2]

```
Method 1: p(X) = [                    8.9E-001,                    1.7E+000]
Method 2: p(X) = [                    8.9E-001,                    1.6E+000]
Method 3: p(X) = [                    1.0E+000,                    1.5E+000]
Method 4: p(X) = [                    1.1E+000,                    1.3E+000]

Enter X : 0

Method 1: p(X) = [  1.000000000000000E+000,  1.000000000000000E+000]
Method 2: p(X) = [  1.000000000000000E+000,  1.000000000000000E+000]
Method 3: p(X) = [  1.000000000000000E+000,  1.000000000000000E+000]
Method 4: p(X) = [  1.000000000000000E+000,  1.000000000000000E+000]
```

Exercise 18: Calculations for Interval Matrices

Let the interval matrices A and B be given:

$$A = \begin{pmatrix} [1,1] & [0,1] \\ [1,1] & [-1,1] \end{pmatrix}, \quad B = \begin{pmatrix} [-1,2] & [3,4] \\ [2,2] & [-6,-4] \end{pmatrix}.$$

A PASCAL–XSC program should

a) calculate $A + B$, $A - B$, $A \cdot B$,

b) demonstrate (by calculation) that $A \cdot (A \cdot A) \neq (A \cdot A) \cdot A$,

c) demonstrate (by calculation) that $A \cdot (B + A) \not\subseteq A \cdot B + A \cdot A$.

Hint: Use the operators supplied in module MVI_ARI.

Solution:

```
program int_matr (input, output);

{ Exercise 18: Calculations for Interval Matrices }

use i_ari, mvi_ari;

var
  A, B, C : imatrix[1..2,1..2];

begin
  writeln ('Exercise 18: Calculations for Interval Matrices');
  writeln;

  A[1,1]:= 1;
  A[1,2]:= intval(0,1);
  A[2,1]:= 1;
  A[2,2]:= intval(-1,1);

  B[1,1]:= intval(-1,2);
  B[1,2]:= intval(3,4);
  B[2,1]:= 2;
  B[2,2]:= intval(-6,-4);

  writeln ('A = '); writeln;
  writeln (A);
  writeln ('B = '); writeln;
  writeln (B);
  writeln ('A + B = '); writeln;
  writeln (A+B);
```

```
   writeln ('A - B = '); writeln;
   writeln (A-B);
   writeln ('A * B = '); writeln;
   writeln (A*B);
   writeln ('A * (A * A) = '); writeln;
   writeln (A*(A*A));
   writeln ('(A * A) * A = '); writeln;
   writeln ((A*A)*A);
   writeln ('A * (B + A) = '); writeln;
   writeln (A*(B+A));
   writeln ('A * B + A * A = '); writeln;
   writeln (A*B+A*A);
end.
```

Runtime Output:

Exercise 18: Calculations for Interval Matrices

A =

```
[ 1.0E+00, 1.0E+00] [ 0.0E+00, 1.0E+00]
[ 1.0E+00, 1.0E+00] [-1.0E+00, 1.0E+00]
```

B =

```
[-1.0E+00, 2.0E+00] [ 3.0E+00, 4.0E+00]
[ 2.0E+00, 2.0E+00] [-6.0E+00,-4.0E+00]
```

A + B =

```
[ 0.0E+00, 3.0E+00] [ 3.0E+00, 5.0E+00]
[ 3.0E+00, 3.0E+00] [-7.0E+00,-3.0E+00]
```

A - B =

```
[-1.0E+00, 2.0E+00] [-4.0E+00,-2.0E+00]
[-1.0E+00,-1.0E+00] [ 3.0E+00, 7.0E+00]
```

A * B =

```
[-1.0E+00, 4.0E+00] [-3.0E+00, 4.0E+00]
[-3.0E+00, 4.0E+00] [-3.0E+00, 1.0E+01]
```

A * (A * A) =

```
[ 1.0E+00, 4.0E+00] [-2.0E+00, 4.0E+00]
[-1.0E+00, 4.0E+00] [-3.0E+00, 4.0E+00]
```

(A * A) * A =

[0.0E+00, 4.0E+00] [-2.0E+00, 4.0E+00]
[-1.0E+00, 4.0E+00] [-2.0E+00, 4.0E+00]

A * (B + A) =

[0.0E+00, 6.0E+00] [-4.0E+00, 5.0E+00]
[-3.0E+00, 6.0E+00] [-4.0E+00, 1.2E+01]

A * B + A * A =

[0.0E+00, 6.0E+00][-4.0E+00, 6.0E+00]
[-3.0E+00, 6.0E+00][-4.0E+00, 1.2E+01]

Exercise 19: Differentiation Arithmetic

With the help of differentiation arithmetic (see [42]), compute the values of the function

$$f(x) = x \cdot \frac{4+x}{3-x}$$

and the values of its derivative $f'(x)$ in the domain $-4 \le x \le 2$ for the points $x_k = -4 + kh$, $k = 0, \dots, 48$ with $h = 0.125$.

Differentiation arithmetic is an arithmetic for ordered pairs of the form

$$U = (u, u') \quad \text{with} \quad u, u' \in \mathbb{R}.$$

The first component of U contains the value of the function. The second contains the value of the derivative. The rules for the arithmetic are

$$
\begin{aligned}
U + V &= (u, u') + (v, v') &&= (u + v, u' + v') \\
U - V &= (u, u') - (v, v') &&= (u - v, u' - v') \\
U * V &= (u, u') * (v, v') &&= (u * v, u * v' + u' * v) \\
U/V &= (u, u')/(v, v') &&= (u/v, (u' - u/v * v')/v), \quad v \ne 0,
\end{aligned}
$$

where the corresponding differentiation rules have to be used in the second component. The independent variable x and the arbitrary constant c correspond to

$$X = (x, 1) \quad \text{and} \quad C = (c, 0),$$

because $\frac{dx}{dx} = 1$, and $\frac{dc}{dx} = 0$. To use the differentiation arithmetic in a PASCAL–XSC program, declare a type $Derivative\_Type$ as $record$ of two $real$ values. Now, define a function f with parameters and result of type $Derivative\_Type$. The operators $+, -, *, /$ perform differentiation arithmetic. If

$$f(X) = X * ((4, 0) + X)/((3, 0) - X),$$

then an automatic differentiation is done because of

$$f(X) = f((x, 1)) = (f(x), f'(x)).$$

That is, the value of the function and the value of the derivative are automatically and simultaneously calculated. Write a PASCAL–XSC module which contains

a) the type declaration $Derivative\_Type$ and

b) the declarations for the operators $+, -, *, /$ according to the rules for the differential arithmetic given above.

Write a PASCAL–XSC program which contains

a) a function F, using the operators of the module and thus delivering the value of the function f and the automatically calculated value of its derivative and

b) a main program that calculates and tabulates the values of $f(x)$ and $f'(x)$ at the specified points.

Hint: The constants 4 and 3 of type *Derivative_Type* in the function f are represented by $(4,0)$ and $(3,0)$, respectively. The independent variable x is represented as $(x,1)$.

Solution:

```
module diff_ari;

{ Exercise 19: Module Providing Differentiation Arithmetic }

global type Derivative_Type = global record
                                    f, df : real;
                                  end;

global operator := (var a: Derivative_Type; r: real);

  begin
    a.f  := r;
    a.df := 0;
  end;

global operator + (a,b: Derivative_Type) Result_add: Derivative_Type;

  begin
    Result_add.f  := a.f + b.f;
    Result_add.df := a.df + b.df;
  end;

global operator - (a,b: Derivative_Type) Result_sub: Derivative_Type;

  begin
    Result_sub.f  := a.f - b.f;
    Result_sub.df := a.df - b.df;
  end;

global operator * (a,b: Derivative_Type) Result_mul: Derivative_Type;

  begin
    Result_mul.f  := a.f * b.f;
    Result_mul.df := a.f * b.df + a.df * b.f;
  end;
```

```
global operator / (a,b: Derivative_Type) Result_div: Derivative_Type;

  begin
    Result_div.f  := a.f / b.f;
    Result_div.df := (a.df - a.f * b.df / b.f) / b.f ;
  end;

end.



program Automatic_Differentiation (input, output);

{ Exercise 19: Differentiation Arithmetic }

use diff_ari;

function f (x: Derivative_Type): Derivative_Type;

  var
    three, four : Derivative_Type;

  begin
    three := 3;
    four  := 4;
    f:= x*((four+x)/(three-x));
  end;

  var
    x, y : Derivative_Type;
    h    : real;
    i    : integer;

begin
  writeln ('Exercise 19: Differentiation Arithmetic');
  writeln;
  x.df := 1;
  h    := 0.125;
  writeln('            x            ', ' ', '  ',
        '              f(x)        ', ' ', '  ',
        '              f''(x)      ');
  for i:= 0 to 48 do
  begin
    x.f := -4 + i * h;
    y   := f(x);
    writeln (x.f,'    ',y.f,'    ',y.df);
  end;
end.
```

Runtime Output:

Exercise 19: Differentiation Arithmetic

x	f(x)	f'(x)
-4.000000000000000E+000	0.000000000000000E+000	-5.714285714285714E-001
-3.875000000000000E+000	-7.045454545454545E-002	-5.557024793388429E-001
-3.750000000000000E+000	-1.388888888888889E-001	-5.390946502057612E-001
-3.625000000000000E+000	-2.051886792452830E-001	-5.215379138483447E-001
-3.500000000000000E+000	-2.692307692307693E-001	-5.029585798816567E-001
-3.375000000000000E+000	-3.308823529411765E-001	-4.832756632064591E-001
-3.250000000000000E+000	-3.900000000000000E-001	-4.624000000000000E-001
-3.125000000000000E+000	-4.464285714285714E-001	-4.402332361516035E-001
-3.000000000000000E+000	-5.000000000000000E-001	-4.166666666666667E-001
-2.875000000000000E+000	-5.505319148936170E-001	-3.915799004074241E-001
-2.750000000000000E+000	-5.978260869565217E-001	-3.648393194706995E-001
-2.625000000000000E+000	-6.416666666666666E-001	-3.362962962962963E-001
-2.500000000000000E+000	-6.818181818181818E-001	-3.057851239669422E-001
-2.375000000000000E+000	-7.180232558139534E-001	-2.731206057328286E-001
-2.250000000000000E+000	-7.500000000000000E-001	-2.380952380952381E-001
-2.125000000000000E+000	-7.774390243902439E-001	-2.004759071980964E-001
-2.000000000000000E+000	-8.000000000000000E-001	-1.599999999999999E-001
-1.875000000000000E+000	-8.173076923076923E-001	-1.163708086785009E-001
-1.750000000000000E+000	-8.289473684210525E-001	-6.925207756232682E-002
-1.625000000000000E+000	-8.344594594594594E-001	-1.826150474799126E-002
-1.500000000000000E+000	-8.333333333333334E-001	3.703703703703709E-002
-1.375000000000000E+000	-8.250000000000000E-001	9.714285714285709E-002
-1.250000000000000E+000	-8.088235294117647E-001	1.626297577854672E-001
-1.125000000000000E+000	-7.840909090909092E-001	2.341597796143251E-001
-1.000000000000000E+000	-7.500000000000000E-001	3.125000000000000E-001
-8.750000000000000E-001	-7.056451612903225E-001	3.985431841831426E-001
-7.500000000000000E-001	-6.500000000000000E-001	4.933333333333334E-001
-6.250000000000000E-001	-5.818965517241379E-001	5.980975029726516E-001
-5.000000000000000E-001	-5.000000000000000E-001	7.142857142857143E-001
-3.750000000000000E-001	-4.027777777777778E-001	8.436213991769548E-001
-2.500000000000000E-001	-2.884615384615384E-001	9.881656804733727E-001
-1.250000000000000E-001	-1.550000000000000E-001	1.150400000000000E+000
0.000000000000000E+000	0.000000000000000E+000	1.333333333333333E+000
1.250000000000000E-001	1.793478260869565E-001	1.540642722117202E+000
2.500000000000000E-001	3.863636363636364E-001	1.776859504132231E+000
3.750000000000000E-001	6.250000000000000E-001	2.047619047619048E+000
5.000000000000000E-001	9.000000000000000E-001	2.360000000000000E+000
6.250000000000000E-001	1.217105263157895E+000	2.722991689750693E+000
7.500000000000000E-001	1.583333333333333E+000	3.148148148148148E+000
8.750000000000000E-001	2.007352941176471E+000	3.650519031141868E+000
1.000000000000000E+000	2.500000000000000E+000	4.250000000000000E+000
1.125000000000000E+000	3.075000000000000E+000	4.973333333333333E+000
1.250000000000000E+000	3.750000000000000E+000	5.857142857142857E+000
1.375000000000000E+000	4.548076923076922E+000	6.952662721893491E+000
1.500000000000000E+000	5.500000000000000E+000	8.333333333333332E+000
1.625000000000000E+000	6.647727272727272E+000	1.010743801652893E+001
1.750000000000000E+000	8.049999999999999E+000	1.244000000000000E+001
1.875000000000000E+000	9.791666666666666E+000	1.559259259259259E+001
2.000000000000000E+000	1.200000000000000E+001	2.000000000000000E+001

Exercise 20: Newton's Method with Automatic Differentiation

The zero of a function $f(x)$ may be computed by Newton's method and a feasible x_0:

$$x_{n+1} = x_n - \frac{f(x_n)}{f'(x_n)}, \quad n = 0, 1, 2, \ldots$$

Using the module defined in the last exercise, the values of the function f and the values of its derivative can be computed simultaneously by using the corresponding operators within the function f.

Write a PASCAL–XSC program to implement Newton's method with the help of differential arithmetic. Use the function

$$f(x) = e^x - x - 5$$

to test your program. Your program should read the starting value x_0, compute five Newton iterations, and print the values x_i and $f(x_i)$ at each iteration.

Test your program with the starting values $x_0 = 2.0$, and $x_0 = -5.0$. The zeros lie at $x = 1.9368470722$ and $x = -4.99321618865$.

Hint: The function e^U can be implemented for the variable $U = (u, u')$ of type *Derivative_Type* by

$$e^U = e^{(u,u')} = (e^u, u' * e^u).$$

Solution:

```
program newt_diff (input, output);

{ Exercise 20: Newton's Method with Automatic Differentiation }

use diff_ari;

function exp (x: Derivative_Type) : Derivative_Type;
  begin
    exp.f := exp(x.f);
    exp.df:= x.df * exp(x.f);
  end;

function f (x: Derivative_Type) : Derivative_Type;
  var
    five : Derivative_Type;
  begin
    five := 5;
    f    := exp(x) - x - five;
  end;
```

```
var
  x, y : Derivative_Type;
  i    : integer;

begin
  writeln ('Exercise 20: Newton''s Method with');
  writeln ('                 Automatic Differentiation');
  writeln;
  write ('Enter starting value x0 : '); read (x.f);
  x.df := 1;
  for i:= 1 to 5 do
  begin
    y    := f(x);
    x.f := x.f - y.f/y.df;
    writeln ('x',i:1,' : ',x.f);
  end;
end.
```

Runtime Output:

```
Exercise 20: Newton's Method with
                 Automatic Differentiation

Enter starting value x0 : 2.0
x1 :   1.939105856497994E+000
x2 :   1.936850383714592E+000
x3 :   1.936847407225395E+000
x4 :   1.936847407220219E+000
x5 :   1.936847407220219E+000

Exercise 20: Newton's Method with
                 Automatic Differentiation

Enter starting value x0 : -5.0
x1 : -4.993216345093695E+000
x2 : -4.993216188647903E+000
x3 : -4.993216188647903E+000
x4 : -4.993216188647903E+000
x5 : -4.993216188647903E+000
```

Exercise 21: Measurement of Time

The time of a clock with the portions hours (h), minutes (m), and seconds (s) between 00.00.00 o'clock and 23.59.59 o'clock is to be represented by a *record* type *Clock*. Write a PASCAL–XSC declaration for an operator + that adds two such times and, depending on the input, subtracts 24 h, so that the result is again representable in type *Clock*. Use this declaration in a program that reads as many as *maxint* times, adds them to a total time, and prints each time subtotal. The input loop is terminated by the input of 0.00.00.

The input of the times should be given as a *string* in the form hh.mm.ss. This input *string* should be converted to the type *Clock* using the *string* functions of PASCAL–XSC. The opposite is done with the total time for the output, i.e. after conversion from type *Clock* to *string*, the time is printed in the form hh.mm.ss.

Solution:

```
program times (input, output);

{ Exercise 21: Measurement of Time }

type
  Clock = record
            hours   : 0..23;
            minutes : 0..59;
            seconds : 0..59;
          end;

var
  a, b         : string[8];
  time, total  : Clock;
  i, p         : integer;

operator + (a, b: Clock) sum : Clock;

  var
    help : 0..119;

  begin
    help        := a.seconds + b.seconds;
    sum.seconds:= help mod 60;
    help        := a.minutes + b.minutes + help div 60;
    sum.minutes:= help mod 60;
    help        := a.hours + b.hours + help div 60;
    sum.hours   := help mod 24;
  end;
```

```
begin
  writeln ('Exercise 21: Measurement of Time');
  writeln;
  total.hours   := 0;
  total.minutes := 0;
  total.seconds := 0;
  i:= 0;
  repeat
    i:= succ (i);
    write ('Please enter difference of time: ');
    readln;
    read (a);
    p              := pos ('.', a);
    time.hours     := ival(substring(a,1,p-1));
    a              := substring(a,p+1,8);
    p              := pos ('.', a);
    time.minutes := ival(substring(a,1,p-1));
    a              := substring(a,p+1,8);
    time.seconds := ival(substring(a,1,8));
    total          := total + time;
    b              := image (total.hours,2)   + '.' +
                      image (total.minutes,2) + '.' +
                      image (total.seconds,2);
    writeln ('New time                     : ', b);
  until (i=maxint) or (time.hours+time.minutes+time.seconds=0);
end.
```

Exercise 22: Iterative Method

Consider the vector iteration method

$$(*) \quad x^{(k+1)} = c + Ax^{(k)}, \quad k = 0, 1, 2, \ldots$$

with $c, x^{(k)} \in \mathbb{R}^n$, $k = 0, 1, 2, \ldots$ and $A \in \mathbb{R}^{n \times n}$. Assume that the spectral radius of A is less than 1 so that the iteration is convergent.

Write a PASCAL–XSC program that implements this method. Design a module *MatVec* that makes available the necessary types, operators, and procedures. This module should contain the following parts:

a) a dynamic type declaration *vector* defined as a one-dimensional array with component type *real*,

b) a dynamic type declaration *matrix* defined as a two-dimensional array with component type *real*,

c) an equality operator = for the comparison of two vectors $a = (a_i)$ and $b = (b_i)$ according to

$$a = b \quad \Longleftrightarrow \quad a_i = b_i, \quad \text{for all } i,$$

d) an operator + for the addition of two vectors $a = (a_i)$ and $b = (b_i)$ according to

$$c = a + b \quad \text{with} \quad c_i = a_i + b_i, \quad \text{for all } i,$$

e) an operator * for the multiplication of a matrix $A = (a_{ij})$ with a vector $x = (x_i)$ according to

$$y = A * x \quad \text{where} \quad y_i = \sum_j a_{ij} x_j, \quad \text{for all } i,$$

by use of the datatype *dotprecision* to ensure that the computation of y_i is done with only one rounding,

f) an overloading of the procedure *read* for the input of a vector,

g) an overloading of the procedure *read* for the input of a matrix,

h) an overloading of the procedure *write* for the output of a vector.

Implement a program *Iteration* which uses the module *MatVec* and contains the following parts:

1) A procedure *Main* with formal parameter n that works with the types, procedures, and operators of module *MatVec*. *Main* should declare the variables c, A, and $x^{(0)}$ necessary for the iteration as vectors (or matrices) with index range $1, \ldots, n$, and read these variables. Furthermore, *Main* should process the iteration $(*)$ until $x^{(k+1)} = x^{(k)}$ or until $k = 20$. Finally, the result vector $x^{(k+1)}$ from the final iteration should be printed.

2) A main program should read the dimension n and call the procedure *main*.

Solution:

```
module MatVec;

{ Exercise 22: Module with Matrix/Vector Operations }

  global type
    vector = dynamic array [*] of real;
    matrix = dynamic array [*,*] of real;

  global operator = (a, b: vector) equ: boolean;
    { Corresponding index ranges of a and b are assumed }
    var
      i : integer;
    begin
      i:= lbound(a) - 1;
      repeat
        i:= i + 1;
      until (a[i] <> b[i]) or (i = ubound(a));
      equ:= (a[i] = b[i]);
    end;

  global operator + (a,b: vector) vadd: vector[lbound(a)..ubound(a)];
    { Corresponding index ranges of a and b are assumed }
    var
      i : integer;
    begin
      for i:= lbound(a) to ubound(a) do
        vadd[i] := a[i] + b[i];
    end;

  global operator * (A: matrix; x: vector)
                        mvmul: vector[lbound(x)..ubound(x)];
    { Corresponding index ranges of A and x are assumed }
    var
      i, j : integer;
      d    : dotprecision;
    begin
      for i:= lbound(A) to ubound(A) do
      begin
        d:= #(0);
        for j:= lbound(A,2) to ubound(A,2) do
          d:= #(d + a[i,j] * x[j]);
        mvmul[i]:= #*(d);
      end;
    end;
```

```
global procedure read (var f: text; var c: vector);
  var
    i : integer;
  begin
    for i:= lbound(c) to ubound(c) do
      read (f, c[i]);
  end;

global procedure read (var f: text; var A: matrix);
  var
    i, j : integer;
  begin
    for i:= lbound(A) to ubound(A) do
      for j := lbound(A,2) to ubound(A,2) do
        read(f, A[i,j]);
  end;

global procedure write (var f: text; c: vector);
  var
    i : integer;
  begin
    for i:= lbound(c) to ubound(c) do
      writeln (f, c[i]);
  end;

end. {module MatVec}



program iterate (input, output);

  { Exercise 22: Iterative Method }

  use matvec;

  var
    n : integer;

  procedure main (n: integer);

    var
      i, j, k               : integer;
      c, x_k, x_k_plus_1, y : vector[1..n];
      A                     : matrix[1..n,1..n];
```

```
  begin
    writeln ('Enter vector c');
    read (c);
    writeln ('Enter matrix A');
    read (A);
    writeln ('Enter vector x0');
    read (x_k_plus_1);

    {Iteration}
    k:= -1;
    repeat
      x_k         := x_k_plus_1;
      k           := k + 1;
      x_k_plus_1:= c + A * x_k;
    until (x_k_plus_1 = x_k) or (k = 20);
    writeln ('Last iterate: ');
    write (x_k_plus_1);
  end;

begin {Main program}
  writeln ('Exercise 22: Iterative Method x_k+1 := c + A x_k');
  writeln;
  write ('Dimension of vectors and matrices? ');
  read (n);
  main (n);
end.
```

Remark: Module *MatVec* is a prototype of the predefined module MV_ARI of
PASCAL–XSC, which uses the predefined types *rvector* and *rmatrix*. Note
that our prototype module does not check the matching of the index ranges.

Exercise 23: Trace of a Product Matrix

The trace of a $n \times n$ matrix $A = (a_{ij})$ is defined by

$$Trace\,(A) := \sum_{i=1}^{n} a_{ii} = a_{11} + \cdots + a_{nn},$$

i.e. the sum of the diagonal elements. Write a PASCAL–XSC program that accepts the dimension n and the two corresponding $n \times n$ matrices A and B, computes the trace of the product matrix $C = A \cdot B$, and prints the value.

Use the module MV_ARI which declares the procedures and operators for the dynamic types *rvector* and *rmatrix*. Implement a function *Trace1* that determines the trace of the product of two matrices with usual arithmetic operations, and a function *Trace2* that uses an #-expression with the same operations to computate the trace with maximum accuracy. Furthermore, implement a function *Trace3* that does the summation in the usual way, but uses a scalar product for the calculation of the diagonal elements of the product matrix. Finally, implement a function *Trace4* that uses an #-expression for the summation used in *Trace3*. Compare the four versions by means of some examples. Test your program also using the matrices

$$A = \begin{pmatrix} 10^9 & 8 & 126 & -237 \\ 100 & 2 & -12 & 1 \\ 10^5 & 10 & -10^7 & 81 \\ 13 & -3 & 30 & 10^{-7} \end{pmatrix}, \quad B = \begin{pmatrix} 10^8 & 85 & 8 & 6 \\ 12 & 3 & 10^3 & 156 \\ 3 & 14 & 10^{10} & 13 \\ 2 & -8332 & -10^4 & -10^{-8} \end{pmatrix}.$$

Solution:

```
program trace (input, output);

{ Exercise 23: Trace of a Product Matrix }

use mv_ari;

var n: integer;

function trace1 (a, b: rmatrix): real;
  var
    i, j : integer;
    s    : real;
  begin
    s:= 0;
    for i:= lbound(a,1) to ubound(a,1) do
      for j:= lbound(a,2) to ubound(a,2) do
        s:= s + a[i,j] * b[j,i];
    trace1:= s;
  end;
```

```
function trace2 (a, b: rmatrix): real;
  var
    i, j : integer;
  begin
    trace2:= #* ( for i:= lbound(a,1) to ubound(a,1) sum
                    ( for j:= lbound(a,2) to ubound(a,2) sum
                        ( a[i,j]*b[j,i] ) ) );
  end;

function trace3 (a, b: rmatrix): real;
  var
    i : integer;
    s : real;
  begin
    s:= 0;
    for i:= lbound(a,1) to ubound(a,1) do
      s:= s + a[i] * rvector(b[*,i]);
    trace3:= s;
  end;

function trace4 (a, b: rmatrix): real;
  var
    i : integer;
  begin
    trace4:= #* ( for i:= lb(a,1) to ub(a,1) sum
                    ( a[i] * rvector (b[*,i]) ) );
  end;

procedure main (n: integer);

  var
    a, b                  : rmatrix[1..n,1..n];
    tr1, tr2, tr3, tr4 : real;

  begin
    writeln('Enter matrix A:');
    read (A);
    writeln('Enter matrix B:');
    read (B);
    tr1:= trace1 (A,B);
    tr2:= trace2 (A,B);
    tr3:= trace3 (A,B);
    tr4:= trace4 (A,B);
    writeln('Trace of A*B computed conventionally       : ',tr1);
    writeln('and with corresponding #-expression        : ',tr2);
    writeln('Trace of A*B computed with scalar product : ',tr3);
    writeln('and with corresponding #-expression        : ',tr4);
  end;
```

```
begin {program trace}
  writeln ('Exercise 23: Trace of a Product Matrix');
  writeln;
  write('Enter dimension of the matrices: ');
  read (n);
  main(n);
end.  {program trace}
```

Runtime Output:

Exercise 23: Trace of a Product Matrix

Enter dimension of the matrices: 4

Enter matrix A:

1e9	8	126	-237
100	2	-12	1
1e5	10	-1e7	81
13	-3	30	1e-7

Enter matrix B:

1e8	85	8	6
12	3	1e3	156
3	14	1e10	13
2	-8332	-1e4	-1e-8

Trace of A*B computed conventionally	:	-1.600000000000000E+001
and with corresponding #-expression	:	5.999999999999999E+000
Trace of A*B computed with scalar product	:	-9.999999999999999E-016
and with corresponding #-expression	:	5.999999999999999E+000

Exercise 24: Calculator for Polynomials

Write a PASCAL–XSC program that provides a calculator for the addition and multiplication of polynomials with *real* coefficients. The degree n of the polynomials should be no more than 5. For two polynomials p and q of degree n with

$$p(x) = \sum_{i=0}^{n} a_i x^i \quad \text{and} \quad q(x) = \sum_{i=0}^{n} b_i x^i,$$

the sum s is defined by

$$s(x) = p(x) + q(x) = \sum_{i=0}^{n} (a_i + b_i) x^i,$$

and the product r by

$$r(x) = p(x) \cdot q(x) = \sum_{i=0}^{n} \sum_{j=0}^{n} a_i b_j x^{i+j}.$$

A module should be written containing

a) a dynamic type definition *Polynomial* where a polynomial is defined as a dynamic *real* array,

b) a procedure to read the coefficients of a polynomial,

c) an operator + with two operands of type *Polynomial* and with a resulting polynomial of the same degree as the operands,

d) an operator * with two operands of type *Polynomial* and with a resulting polynomial of appropriate degree, implemented with maximum accuracy,

e) a procedure for the output of polynomials.

A program testing this module should contain a procedure *Main* with parameter n (≤ 5), that declares the three polynomials (p, q, s) of degree n and a polynomial (r) of degree $2n$, and reads p and q. Depending upon the user's input, the procedure should compute and print the sum s or the product r. In the main program of this test program, only the degree of the polynomials should be entered and the procedure *Main* be called.

Solution:

```
module poly;

{ Exercise 24: Calculator for Polynomials }

global type Polynomial = dynamic array [*] of real;
```

```
global procedure read (var f: text; var a: Polynomial);
  var i: integer;
  begin
    for i:= 0 to ub(a) do
      read (f, a[i]);
  end;

global operator + (a, b: Polynomial) Result_Add: Polynomial[0..ub(a)];
  var i: integer;
  begin
    for i:= 0 to ub(a) do
      Result_Add[i]:= a[i] + b[i];
  end;

global operator * (a, b: Polynomial) Result_Mul: Polynomial[0..2*ub(a)];
  var i, j, n: integer;
  begin
    n:= ub(a);
    for i:= 0 to n do
      Result_Mul[i]:= #* (for j:= 0   to i  sum (a[j] * b[i-j]));
    for i:= n+1 to 2*n do
      Result_Mul[i]:= #* (for j:= i-n to n  sum (a[j] * b[i-j]));
  end;

global procedure write (var f: text; a: Polynomial);
  var i: integer;
  begin
    write (f, a[0], '    ');
    for i:= 1 to ub(a) do
    begin
      writeln(f, ' + ');
      write('     ', a[i], ' x^' ,i:1);
    end;
    writeln(f);
  end;

end. {module poly}

program test_poly (input, output);

use poly;

var  n, option : integer;

procedure Main (n : integer; var option : integer);
  var
    p, q, s : Polynomial[0..n];
    r       : Polynomial[0..2*n];
```

```
  begin
    writeln('Enter the coefficients of p (0 to n):');
    read (p);
    writeln;
    writeln('Enter the coefficients of q (0 to n):');
    read (q);
    writeln;
    repeat
      writeln('Please select:');
      writeln(' p + q                 ==> 0');
      writeln(' p * q                 ==> 1');
      writeln(' New polynomials p,q ==> 2');
      writeln(' Terminate program   ==> 9');
      writeln;
      write  ('Selection ==> '); read(option);
      writeln;
      if option = 0 then
      begin
        s:= p+q;
        write('p   = '); writeln(p);
        write('q   = '); writeln(q);
        write('p+q = '); writeln(s);
      end
      else if option = 1 then
      begin
        r:= p*q;
        write('p   = '); writeln(p);
        write('q   = '); writeln(q);
        write('p*q = '); writeln(r);
      end;
    until (option <> 0) and (option <> 1);
    writeln;
  end;

begin { test_poly }
  writeln('Exercise 24: Calculator for Polynomials');
  writeln;
  repeat
    repeat
      write('Degree n of the polynomials (>= 0 and <= 5) : ');
      read (n);
    until (0<=n) and (n<=5);
    writeln;
    Main (n, option);
  until (option = 9);
end.
```

Runtime Output:

Exercise 24: Calculator for Polynomials

Degree n of the polynomials (>= 0 and <= 5) : 4

Enter the coefficients of p (0 to n):
99 11 22 33 44

Enter the coefficients of q:
0 1 2 3 4

Please select:
 p + q   ==> 0
 p * q   ==> 1
 New polynomials p,q ==> 2
 Terminate program   ==> 9

Selection ==> 0

p   =   9.900000000000000E+001   +
        1.100000000000000E+001 x^1 +
        2.200000000000000E+001 x^2 +
        3.300000000000000E+001 x^3 +
        4.400000000000000E+001 x^4

q   =   0.000000000000000E+000   +
        1.000000000000000E+000 x^1 +
        2.000000000000000E+000 x^2 +
        3.000000000000000E+000 x^3 +
        4.000000000000000E+000 x^4

p+q =   9.900000000000000E+001   +
        1.200000000000000E+001 x^1 +
        2.400000000000000E+001 x^2 +
        3.600000000000000E+001 x^3 +
        4.800000000000000E+001 x^4

Please select:
 p + q   ==> 0
 p * q   ==> 1
 New polynomials p,q ==> 2
 Terminate program   ==> 9

Selection ==> 1

```
p   =  9.900000000000000E+001      +
       1.100000000000000E+001  x^1 +
       2.200000000000000E+001  x^2 +
       3.300000000000000E+001  x^3 +
       4.400000000000000E+001  x^4


q   =  0.000000000000000E+000      +
       1.000000000000000E+000  x^1 +
       2.000000000000000E+000  x^2 +
       3.000000000000000E+000  x^3 +
       4.000000000000000E+000  x^4


p*q =  0.000000000000000E+000      +
       9.900000000000000E+001  x^1 +
       2.090000000000000E+002  x^2 +
       3.410000000000000E+002  x^3 +
       5.060000000000000E+002  x^4 +
       2.200000000000000E+002  x^5 +
       2.750000000000000E+002  x^6 +
       2.640000000000000E+002  x^7 +
       1.760000000000000E+002  x^8


Please select:
  p + q                ==> 0
  p * q                ==> 1
  New polynomials p,q  ==> 2
  Terminate program    ==> 9

Selection  ==> 9
```

Exercise 25: Interval Newton Method

The interval inclusion X_n of a zero of a function $f(x)$ whose derivative is continuous and not equal to zero on the interval $[a, b]$ can be improved under the assumption $f(a) \cdot f(b) < 0$ with help of the interval Newton method ([1],[2],[34])

$$X_{n+1} := \left(m(X_n) - \frac{f(m(X_n))}{f'(X_n)} \right) \cap X_n.$$

$m(X)$ is the midpoint of the interval X.

Write a PASCAL–XSC program that uses the module I_ARI and computes the interval inclusion of the zero of

$$f(x) = \sqrt{x} + (x + 1) \cos x$$

with the method described above. Your program should include

- a function F that computes $f(X)$ with interval arithmetic

- a function DF that computes the derivative $f'(x)$ with interval arithmetic

- a function *midpoint* that computes the midpoint m of the interval $X = [x_1, x_2]$ with maximum accuracy by means of an #-expression

- a main program that accepts the starting interval $X = [a, b]$, checks the two criteria $f(a) \cdot f(b) < 0$ and $0 \notin DF(X)$, and computes the iterates using the Newton method. Print the newly calculated interval at each iteration. The iteration should terminate when $X_{n+1} = X_n$ (for the finite convergence see [34]).

Hint: Use $[2.0, 3.0]$ as the starting interval for the iteration. Note, that for the computation of $f(m(X))$ with the interval function F, the midpoint delivered by *midpoint* must be converted in an interval.

Solution:

```
program i_newton (input, output);

{ Exercise 25: Interval Newton Method }

use i_ari;

var  x, y : interval;

function F (x : interval) : interval;
  begin
    F:= sqrt(x) + (x+1) * cos(x);
  end;
```

```
function DF (x : interval) : interval;
  begin
    DF:= 0.5/sqrt(x) + cos(x) - (x + 1) * sin(x);
  end;

function midpoint (x : interval) : real;
  begin
    midpoint:= #* (0.5 * x.inf + 0.5 * x.sup);
  end;

function criterion_satisfied (x : interval) : boolean;
  var
    a, b: interval;
  begin
    a:= intval(inf(x));
    b:= intval(sup(x));
    criterion_satisfied:= (sup(F(a)*F(b)) < 0) and (not (0 in DF(x)));
  end;

begin
  writeln('Exercise 25: Interval Newton Method'); writeln;
  write ('Starting interval: '); read (y); writeln;
  writeln ('Iteration'); writeln;
  if criterion_satisfied (y) then
    repeat
      writeln (y);
      x:= y;
      y:= ( midpoint(x) - F ( intval(midpoint(x)) ) / DF(x) ) ** x;
    until y = x
  else
    writeln ('Criterion not satisfied!');
end.
```

Runtime Output:

Exercise 25: Interval Newton Method

Starting interval: [2,3]

Iteration

```
[                2.0E+000,                3.0E+000]
[                2.0E+000,                2.3E+000]
[               2.05E+000,               2.07E+000]
[            2.05903E+000,            2.05906E+000]
[      2.059045253413E+000,      2.059045253417E+000]
[  2.059045253415143E+000,  2.059045253415145E+000]
```

Exercise 26: Runge-Kutta Method

The Runga-Kutta method [47] is used for approximating a solution for initial value problems of the form

$$Y' = F(x, Y); \quad Y(x^0) = Y^0;$$

where

$$Y = \begin{pmatrix} y_1(x) \\ \vdots \\ y_n(x) \end{pmatrix}, \quad Y' = \begin{pmatrix} y_1'(x) \\ \vdots \\ y_n'(x) \end{pmatrix}$$

and

$$F(x, Y) = \begin{pmatrix} f_1(x, y_1, \ldots, y_n) \\ \vdots \\ f_n(x, y_1, \ldots, y_n) \end{pmatrix}.$$

Define the coefficients K_i

$$\begin{aligned}
K_1 &= h * F(x, Y) \\
K_2 &= h * F(x + \tfrac{h}{2}, Y + \tfrac{K_1}{2}) \\
K_3 &= h * F(x + \tfrac{h}{2}, Y + \tfrac{K_2}{2}) \\
K_4 &= h * F(x + h, Y + K_3).
\end{aligned}$$

An approximation for the solution Y at the point $x + h$ is given by the formula

$$Y(x + h) = Y(x) + (K_1 + 2K_2 + 2K_3 + K_4)/6.$$

Write a PASCAL–XSC program that uses the module MV_ARI. Starting from

$$Y(0) = \begin{pmatrix} 1 \\ 0 \\ 1 \end{pmatrix},$$ the values of Y at the points $x_i = i * h$, $i = 1, \ldots, 10$ should be computed with $h = 0.125$. As an example, use the function

$$F(x, Y) = \begin{pmatrix} Y_1 - Y_2 \\ e^x \cdot Y_3 \\ (Y_1 - Y_2)/e^x \end{pmatrix}.$$

The output should be presented as a table. Define the vector function $F(x, Y)$. Compute the expressions K_1, K_2, K_3, K_4, and the value of $Y(x_i)$ in a loop using the predefined operators in MV_ARI.

Solution:

```pascal
program Runge_Kutta (input, output);

{ Exercise 26: Runge-Kutta Method }

use mv_ari;

const
  n = 3;

var
  h, x : real;
  Y    : rvector[1..n];
  i    : integer;

function F (x : real; Y : rvector) : rvector[1..n];
  var
    i: integer;
  begin
    F[1]:= Y[1] - Y[2];
    F[2]:= exp(x) * Y[3];
    F[3]:= (Y[1] - Y[2]) / exp(x);
  end;

function One_Step (x, h : real; var Y : rvector) : rvector[1..n];
  { This function executes one step of the Runge-Kutta method }
  var
    k1, k2, k3, k4 : rvector[1..n];
  begin
    k1 := h * F (x        , Y);
    k2 := h * F (x + h/2, Y + k1/2);
    k3 := h * F (x + h/2, Y + k2/2);
    k4 := h * F (x + h  , Y + k3);
    One_Step := Y + (k1 + 2 * k2 + 2 * k3 + k4) / 6;
  end;

begin
  writeln('Exercise 26: Runge-Kutta Method');
  writeln;
  x:= 0;  Y[1]:= 1;  Y[2]:= 0;  Y[3]:= 1;  h:= 0.125;
  writeln ('   x                                              Y');
  write   ('------------------------------------------------');
  writeln ('------------------------------------------------');
  writeln (xi:7:4,'  ',Y[1],'  ',Y[2],'  ',Y[3]);
```

```
  for i:=1 to 10 do
  begin
    x := i*h;
    Y := One_Step (x, h, Y);
    writeln (x:7:4,'   ',Y[1],' ',Y[2],' ',Y[3]);
  end;
end.
```

Runtime Output:

Exercise 26: Runge-Kutta Method

x	Y		
0.0000	1.000000000000000E+000	0.000000000000000E+000	1.000000000000000E+000
0.1250	1.123177059359435E+000	1.589550140041404E-001	1.102222238011605E+000
0.2500	1.239209386091870E+000	3.550710525572459E-001	1.187244664965584E+000
0.3750	1.341958893792722E+000	5.919581434349622E-001	1.253740540716452E+000
0.5000	1.424014481447932E+000	8.728079152996371E-001	1.300672225680391E+000
0.6250	1.476592898728182E+000	1.200161743578859E+000	1.327307372755333E+000
0.7500	1.489460043511421E+000	1.575636069976619E+000	1.333230355256656E+000
0.8750	1.450881281820536E+000	1.999602458523678E+000	1.318348752540877E+000
1.0000	1.347610792483193E+000	2.470820861695647E+000	1.282894792112784E+000
1.1250	1.164931397828429E+000	2.986025793442226E+000	1.227421725713755E+000
1.2500	8.867578092820163E-001	3.539466687473293E+000	1.152795195942419E+000

Remark: The solutions of both of the last exercises demonstrate that the general operator concept in PASCAL–XSC substantially simplifies the transfer of numerical algorithms into program code. In principle, the mathematical formulas can be used directly as program statements.

Exercise 27: Rational Arithmetic

Implement a PASCAL–XSC module for a rational arithmetic [23]. A rational number $p = n/d$ should be represented as a *record* type with *integer* components *numerator* and *denominator* (> 0). The module should make the following globally available

1) the type *Rational*,

2) the operators $+,-,*,/$, which deliver a *reduced* fraction of type *Rational* as result,

3) a procedure for the input and for the output of rational numbers, respectively, using the form:

$$integer/integer$$

You will need to write functions to compute the greatest common denominator (gcd) and to reduce fractions. These should be declared locally for use only within the module.

A test program should test each operator and compute the value of the expression

$$(a + b) * (b - c)/(c + d)$$

for $a = 3/4$, $b = 2/7$, $c = 4/5$, and $d = 7/9$.

Hint: The function to reduce a rational number to lowest terms should use integer division (**div**) of the denominator and the numerator by the greatest common divisor . The function for the greatest common divisor should use the following algorithm:

$$a, b > 0; \quad z_0 := a; \quad n_0 := b; \quad i := 0;$$

$$\text{set} \left\{ \begin{array}{l} n_{i+1} := d_i \\ d_{i+1} := n_i \bmod d_i \end{array} \right\} \text{ for } i = 0, 1, 2, \ldots$$

until $d_{i+1} = 0$.

then n_{i+1} (or d_i) is the greatest common divisor of a and b.

Notice: $\gcd(0, x) = x$ for every $x \neq 0$.

Solution:

```
module rational;

{ Exercise 27: Rational Arithmetic }

global type
  positive = 1..maxint;
  rational = record
                numerator   : integer;
                denominator : positive;
             end;

function gcd (a, b : integer) : positive;
  var
    n, d, r : integer;
  begin
    if a = 0 then
      gcd:= b
    else if b = 0 then
      gcd:= a
    else
      begin
        d:= a;
        r:= b;
        repeat
          n:= d;
          d:= r;
          r:= n mod d;
        until r = 0;
        gcd:= abs(d);
      end;
  end;

function reduce (a: rational) : rational;
  var
    g : positive;
  begin
    g:= gcd (abs(a.numerator),a.denominator);
    if (g = 0) or (g = 1) then
      reduce := a
    else
      begin
        reduce.numerator  := a.numerator   div g;
        reduce.denominator:= a.denominator div g;
      end;
  end;
```

```
global operator + (a,b : rational) respl : rational;
  var
    s: rational;
  begin
    s.numerator  := a.numerator*b.denominator + b.numerator*a.denominator;
    s.denominator:= a.denominator*b.denominator;
    respl:= reduce (s);
  end;

global operator - (a,b : rational) resmi : rational;
  var
    s: rational;
  begin
    s.numerator  := a.numerator*b.denominator - b.numerator*a.denominator;
    s.denominator:= a.denominator*b.denominator;
    resmi:= reduce (s);
  end;

global operator * (a,b : rational) resmu : rational;
  var
    s: rational;
  begin
    s.numerator  := a.numerator*b.numerator;
    s.denominator:= a.denominator*b.denominator;
    resmu:= reduce (s);
  end;

global operator / (a,b : rational) resdi : rational;
  var
    help : integer;
    s    : rational;
  begin
    s.numerator:= a.numerator*b.denominator;
    help       := a.denominator*b.numerator;
    if help > 0 then
      s.denominator:= help
    else if help < 0 then
      begin
        s.numerator  := -s.numerator;
        s.denominator:= -help;
      end
    else    { force division by zero to generate an error }
      help:= help div help;
    resdi:= reduce (s);
  end;
```

```
global procedure read (var f: text; var r: rational);
  var
    s, sn, sd: string;
    i, l     : integer;
  begin
    if eoln (f) then
      readln (f);
    read (f,s);
    i:= pos ('/',s);
    l:= length (s);
    sn:= substring (s,1,i-1);
    sd:= substring (s,i+1,l-i);
    r.numerator:= ival (sn);
    l:= ival (sd);
    if l > 0 then
      r.denominator:= l
    else if l < 0 then
      begin
        r.numerator  := - r.numerator;
        r.denominator:= - l;
      end
    else  { force devision by zero to generate an error }
      l:= l div l;
    r:= reduce (r);
  end;

global procedure write (var f: text; a: rational);
  begin
    write (f, a.numerator:1, '/', a.denominator:1);
  end;

end. {module rational}



program test_ratio (input, output);

{ Exercise 27: Rational Arithmetic - Test Program }

use rational;

var
  a,b,c,d : rational;
```

```
begin
  writeln('Exercise 27: Rational Arithmetic - Test Program');
  writeln;
  write ('a = '); read (a);
  writeln (a);
  write ('b = '); read (b);
  writeln (b);
  write ('c = '); read (c);
  writeln (c);
  write ('d = '); read (d);
  writeln (d);
  writeln;
  writeln ('a+b = ', a+b);
  writeln ('b-c = ', b-c);
  writeln ('c+d = ', c+d);
  writeln ('(a+b)*(b-c)/(c+d) = ', (a+b)*(b-c)/(c+d));
end.
```

Runtime Output:

```
Exercise 27: Rational Arithmetic

a   =     3/4
b   =     2/7
c   =     4/5
d   =     7/9

a+b =   29/28
b-c = -18/35
c+d =   71/45
(a+b)*(b-c)/(c+d) = -2349/6958
```

Exercise 28: Evaluation of Polynomials

Write a PASCAL–XSC program to evaluate a polynomial

$$p(t) = a_n t^n + \cdots + a_1 t + a_0$$

with maximum accuracy. Use the module LSS from the PASCAL–XSC numeric library for the verified solution of a system of linear equations. Horner's scheme

$$p(t) = (\ldots (a_n \cdot t + a_{n-1}) \cdot t + a_{n-2}) \cdots) \cdot t + a_1) \cdot t + a_0$$

for the evaluation of a polynomial can be done via the solution of the system of linear equations

$$
\begin{aligned}
x_0 &= a_n \\
x_1 &= t x_0 + a_{n-1} \\
&\vdots \\
x_{n-1} &= t x_{n-2} + a_1 \\
x_n &= t x_{n-1} + a_0
\end{aligned}
$$

by introducing the $n+1$ variables $x_0, x_1, \ldots, x_{n-1}, x_n$. The value of the polynomial p at point t is then given by x_n, i.e. $x_n = p(t)$.

Hence, we wish to solve the system of linear equations

$$
\begin{pmatrix}
1 & & & & 0 \\
-t & 1 & & & \\
& \ddots & \ddots & & \\
& & -t & 1 & \\
0 & & & -t & 1
\end{pmatrix}
\cdot
\begin{pmatrix}
x_0 \\
x_1 \\
\vdots \\
x_{n-1} \\
x_n
\end{pmatrix}
=
\begin{pmatrix}
a_n \\
a_{n-1} \\
\vdots \\
a_1 \\
a_0
\end{pmatrix}
$$

or

$$Ax = b$$

where

$$A = (a_{ij}), \quad a_{ij} = \begin{cases} 1 & \text{for } i = j \\ -t & \text{for } i = j+1 \\ 0 & \text{else} \end{cases}, \quad i, j = 0, \ldots, n$$

and

$$b = (b_i), \quad b_i = a_{n-i}, \quad i = 0, \ldots, n.$$

Write a PASCAL–XSC program that contains the following parts:

(a) a dynamic type declaration *polynomial* (component type *real*),

(b) a procedure *read* for the coefficients of a polynomial,

(c) a function *Horner* to compute the value of a polynomial by the Horner scheme,

(d) a procedure *set_A_b*, that generates the matrix A and the vector b from a polynomial p and a *real* number t,

(e) a procedure *main* with formal parameter n that

  – declares a variable p of type *polynomial*, a vector b of type *rvector*, an interval vector X of type *ivector*, and a square matrix A of type *rmatrix* with index range $0, \ldots, n$,

  – reads the polynomial coefficients $a_0, \ldots, a_n$ using the procedure *read* of part (b),

  – generates the matrix A and the vector b using the procedure from part (d),

  – computes an inclusion of X of the solution of the system $Ax = b$ with maximum accuracy using the procedure *lss*,

  – and finally, if *lss* is executed without errors, prints the lower and upper bounds of the interval inclusion X_n of the polynomial value $x_n = p(t)$ and the value calculated by the Horner method (part (c)) for the sake of comparison.

(f) a main program that accepts the degree of the polynomial n and calls the procedure *main*.

Hint: Use the module LSS from the PASCAL–XSC numeric library. This module supplies the procedure *lss* which delivers a verified inclusion vector X for the solution x of $Ax = b$ using the matrix A and the right-hand side b as input. The interface of this procedure is

$$\text{procedure lss (} \quad \text{var A: rmatrix;} \quad \text{var b: rvector;}$$
$$\text{var X: ivector;} \quad \text{var errcode: integer });$$

where:

$$\text{errcode} = 0: \quad \text{errorfree execution,}$$
$$\text{errcode} = 1: \quad \text{system is too poorly conditioned,}$$
$$\text{errcode} = 2: \quad \text{matrix is possibly singular.}$$

Test your program with the following examples:

Example 1:

degree of polynomial   3
coefficients
$$a_0 = 1536796802$$
$$a_1 = -1086679440$$
$$a_2 = -768398401$$
$$a_3 = 543339720$$
point of evaluation $t = 1.4142135$

Example 2:

degree of polynomial 3

coefficients

$a_0 = 191971912515$

$a_1 = -135744641136$

$a_2 = -95985956257$

$a_3 = 67872320568$

point of evaluation $t = 1.41421353154$

Solution:

```
program Polynomial_Evaluation (input, output);

{ Exercise 28: Evaluation of Polynomials }

use i_ari, mv_ari, lss;

type  polynomial = dynamic array [*] of real;

procedure read (var f: text; var p: polynomial);
  var
    i: integer;
  begin
    for i:= 0 to ub(p) do
    begin
      write ('Coeff. ',i:2,': ');
      read (f, p[i]);
    end;
  end;

function Horner (p : polynomial; t: real) : real;
  var
    h: real;
    i: integer;
  begin
    h:= 0;
    for i:= ub(p) downto 0 do
      h:= p[i] + t * h;
    horner:= h;
  end;

procedure set_A_b (p    : polynomial; t    : real;
                   var A: rmatrix;    var b: rvector);
  var
    i, j, ub_p: integer;
  begin
    A:= null (A);
    A[0,0]:= 1;
```

```
      for i:= 1 to ub(A) do
      begin
        A[i,i-1]:= -t; { sub diagonal := -t }
        A[i,i]   := 1; { diagonal     := 1 }
      end;
      ub_p:= ub(p);
      for i:= 0 to ub(b) do
        b[i] := p[ub_p-i];
   end;

procedure main (n: integer);
   var
     p: polynomial[0..n];
     b: rvector[0..n];
     X: ivector[0..n];
     A: rmatrix[0..n,0..n];
     t: real;
     error: integer;
   begin
     writeln ('Enter a polynomial');
     read (p);
     write ('Enter the point of evaluation  t = '); read(t);
     writeln;
     set_A_b (p,t,A,b);
     lss (A,b,X,error);
     if error=0 then
       begin
         writeln ('Horner scheme :    ', horner (p,t));
         writeln ('Inclusion     : ',    X[n]);
       end
     else
       writeln ('Error ',error:1,' ocurred');
   end;

var   n: integer;

begin
  writeln('Exercise 28: Evaluation of Polynomials');
  writeln;
  write ('Degree of polynomial: ');   read (n);
  main (n);
end.
```

Runtime Output:

Example 1

```
Degree of polynomial: 3
Enter polynomial
Coeff.  0 =  1536796802
Coeff.  1 = -1086679440
Coeff.  2 = -768398401
Coeff.  3 =  543339720
Enter the point of evaluation  t = 1.4142135
```

```
Horner scheme :    5.960464477539062E-006
Inclusion     : [  5.978758733249328E-006,  5.978758733249330E-006]
```

Example 2

```
Degree of polynomial: 3
Enter polynomial
Coeff.  0 = 191971912515
Coeff.  1 = -135744641136
Coeff.  2 = -95985956257
Coeff.  3 = 67872320568
Enter the point of evaluation  t = 1.41421353154
```

```
Horner scheme :    1.000183105468750E+000
Inclusion     : [  1.000182503810985E+000,  1.000182503810986E+000]
```

Remark: This last exercise shows how to use the routine for the verified solution of linear equations to evaluate poynomials with maximum accuracy. The verified results show that the frequently used Horner method may deliver incorrect results.

The procedure *lss* is used for simplicity. The reader might wish to design and implement a more efficient algorithm which takes advantage of the special structure of the matrix A as an advanced exercise (see [8]).

Appendix A

Syntax Diagrams

As a supplement to the syntax description of the language reference (chapter 2) using the simplified Backus-Naur-Form, we now give a complete description of the PASCAL–XSC syntax. For this purpose, we use syntax diagramms in a special form being already mentioned in [6], [7] or [14]. The following rules apply to the usage of the diagrams.

- Each diagram is marked by a number followed by a special identifier (sequence of upper case letters). This identifier is called a *syntax variable*. It is chosen to refer to the represented language element.

- A diagram consists of syntax variables, terminal symbols (reserved words consisting of **boldfaced** sequences of lower case letters, symbols enclosed in circles, or sequences of symbols enclosed in ovals), and solid or dotted lines.

- Within a diagram, a syntax variable may occur in connection with a semantic prefix. For instance, the variable IDENTIFIER (ID) is used with the prefix *COMP* indicating a special kind of identifier, i.e. a component identifier. Nevertheless, the definition of the variable *COMP* IDENTIFIER is given by the syntax diagram IDENTIFIER.

  Furthermore, these semantic attributes appear as *italicized* remarks which are stated immediately beneath or beside a variable. If a list of semantic attributes is given, then the comma is read as "or". The following abbreviations are used:

A	*Array*
B	*boolean*
CH	*char*
CIR	*cinterval (complex intervals)*
COMP	*Component (of a record)*
CONST	*Constant*
CR	*complex (complex numbers)*
DOT	*dotprecision*
dyadop	*dyadic operator*
dyna	*dynamic array*
ET	*Enumeration Type*
F	*File*

FCT	*Function*
FL	*Field List*
FS	*File Structure Type*
I	*integer*
id	*identifier*
IR	*interval* (real intervals)
MCIR	*cimatrix* (complex interval matrices)
MCR	*cmatrix* (complex matrices)
MIR	*imatrix* (interval matrices)
MR	*rmatrix* (real matrices)
monop	*monadic operator*
P	*pointer*
R	*real*
REC	*Record*
RES	*Result*
ST	*string*
TF	*Text File*
VCIR	*civector* (complex interval vectors)
VCR	*cvector* (complex vectors)
VIR	*ivector* (interval vectors)
VR	*rvector* (real vectors)
VAR	*Variable*

An index of all syntax variables (identifiers) is listed in alphabetical order after the diagrams in Appendix B.1 to simplify working with the diagrams.

While editing a program, the syntax diagrams are used according to the following rules:

- The traversing of a diagram starts at the upper left.
- Solid lines must be followed from left to right or from top to bottom. Dotted lines must be followed from right to left or from bottom to top.
- The traversing of a diagram ends at the lower right.
- Wherever a syntax variable appears while traversing over a diagram, we have to traverse through the diagram of this syntax variable. Then we continue with the original diagram.

P1   COMPILATION UNIT

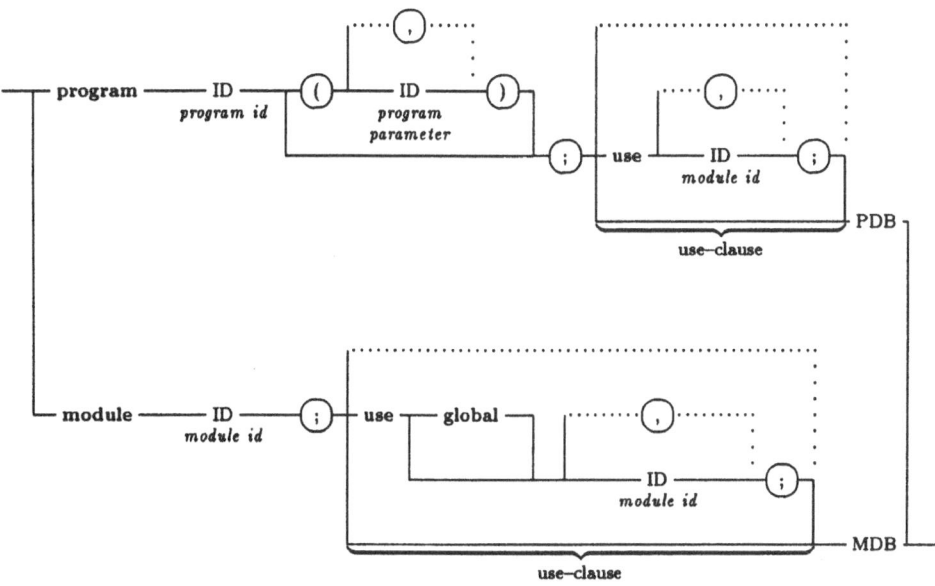

P2   PROGRAM DECLARATION AND BODY   (PDB)

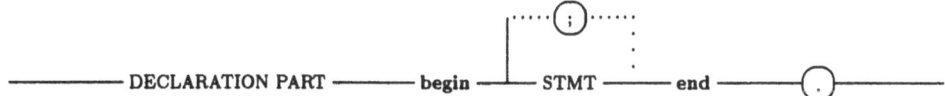

P3   MODULE DECLARATION AND BODY   (MDB)

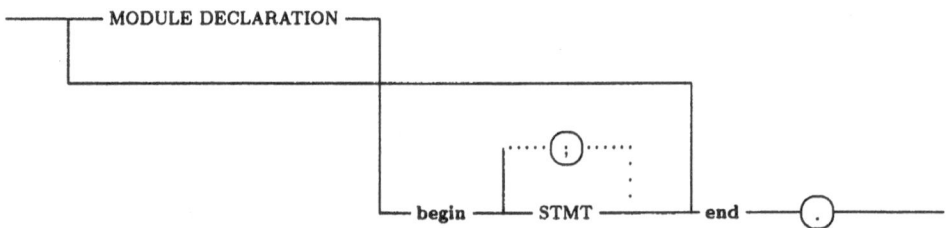

P4   MODULE DECLARATION

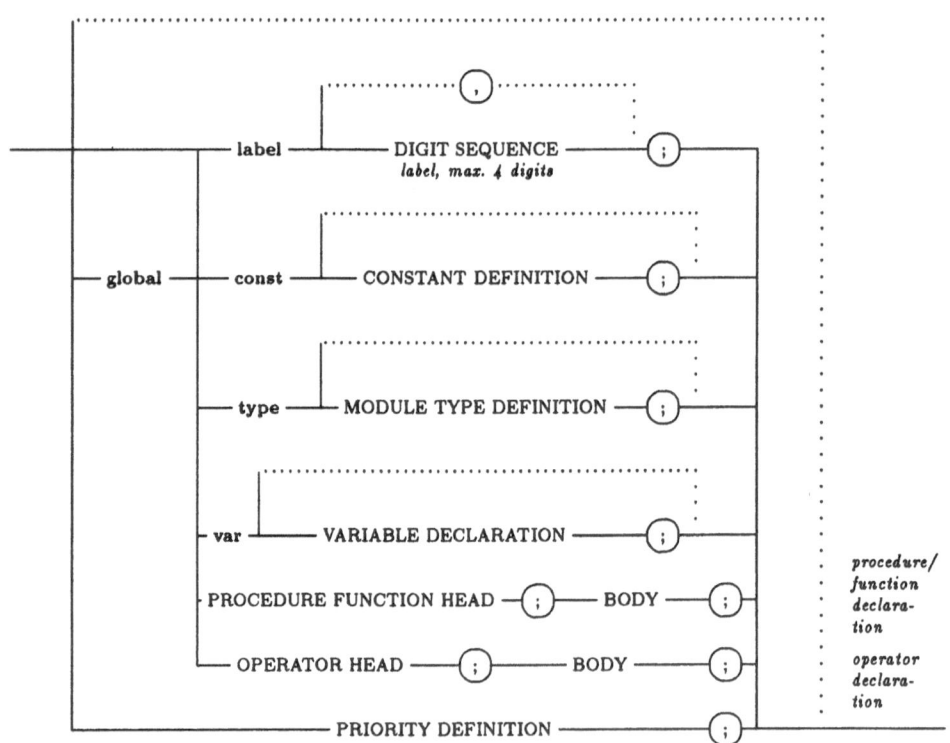

P5   DECLARATION PART

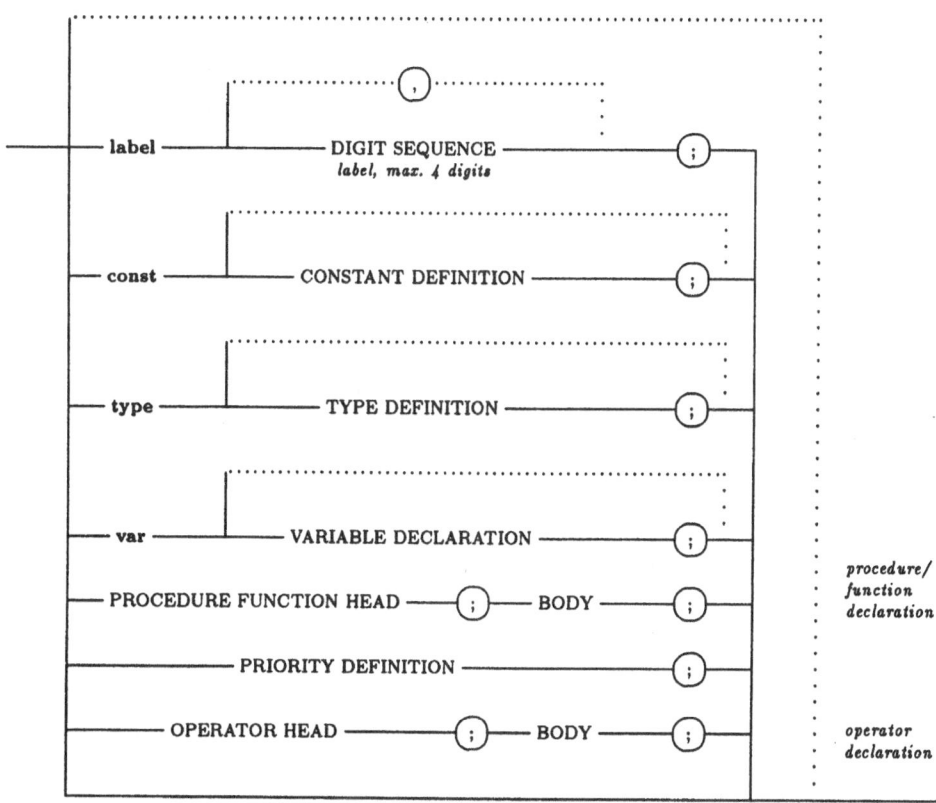

P6   CONSTANT DEFINITION

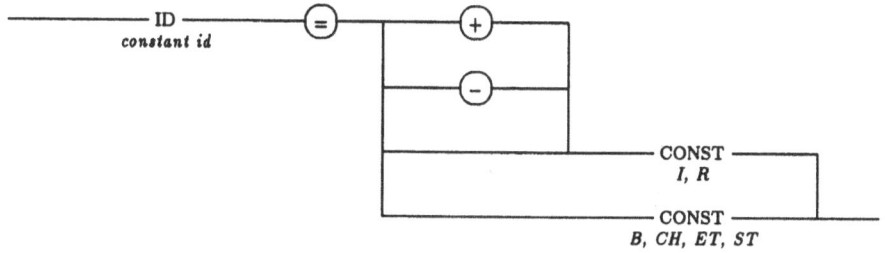

P7   CONSTANT   (CONST)

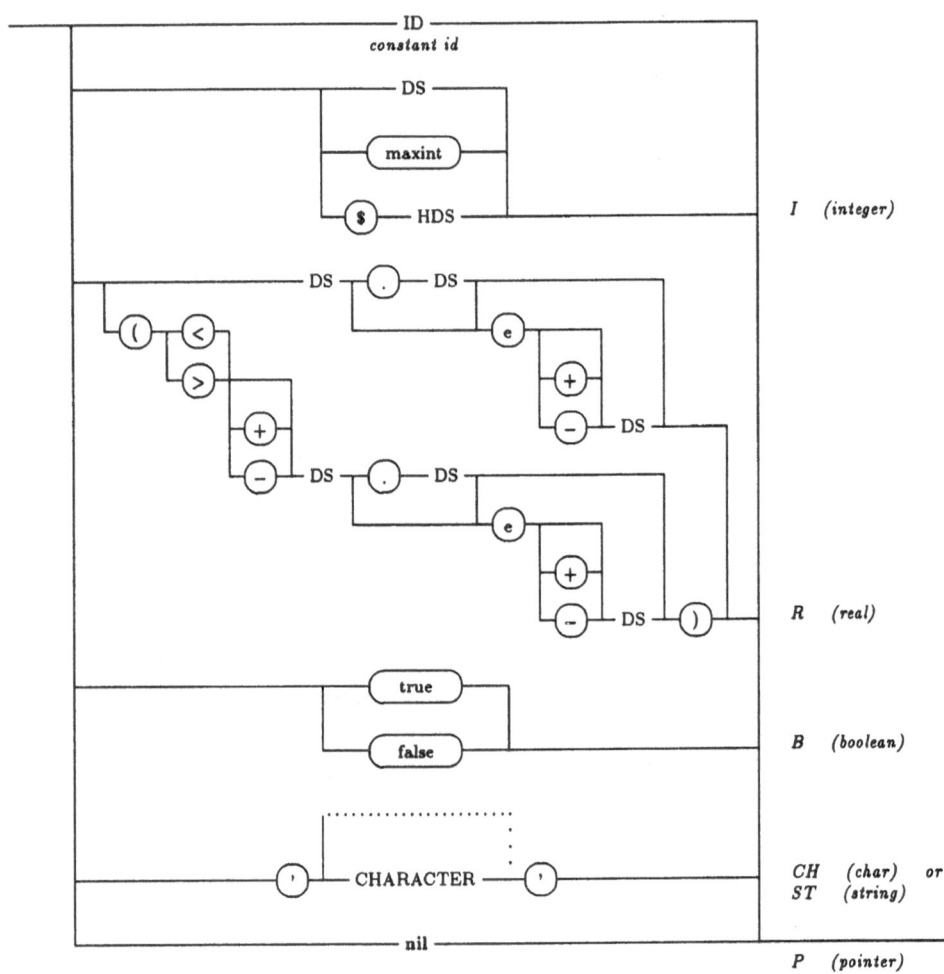

P8   TYPE DEFINITION

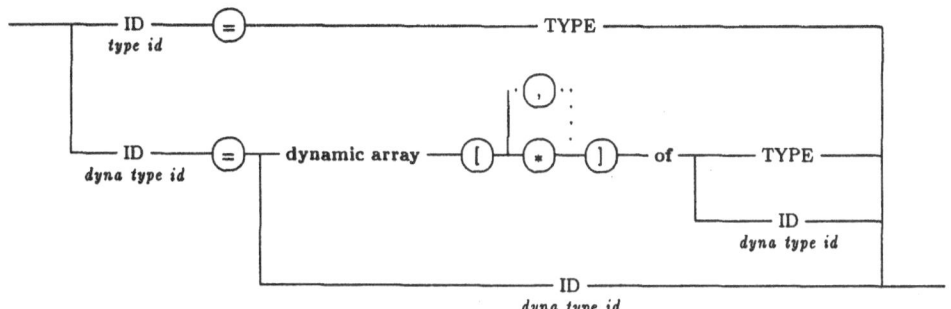

List of predefined PASCAL–XSC type identifiers:

integer	I	dotprecision	DOT
real	R	rvector	VR
boolean	B	rmatrix	MR
char	CH	cvector	VCR
text	TF	cmatrix	MCR
string	ST	ivector	VIR
complex	CR	imatrix	MIR
interval	IR	civector	VCIR
cinterval	CIR	cimatrix	MCIR

P9   MODULE TYPE DEFINITION

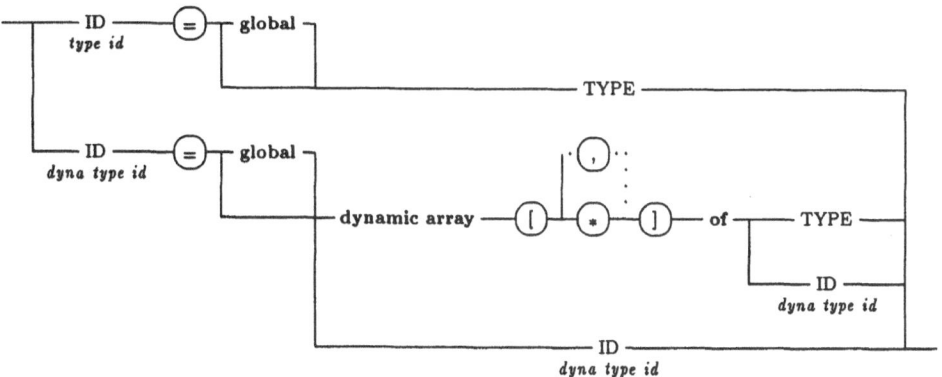

P10   TYPE

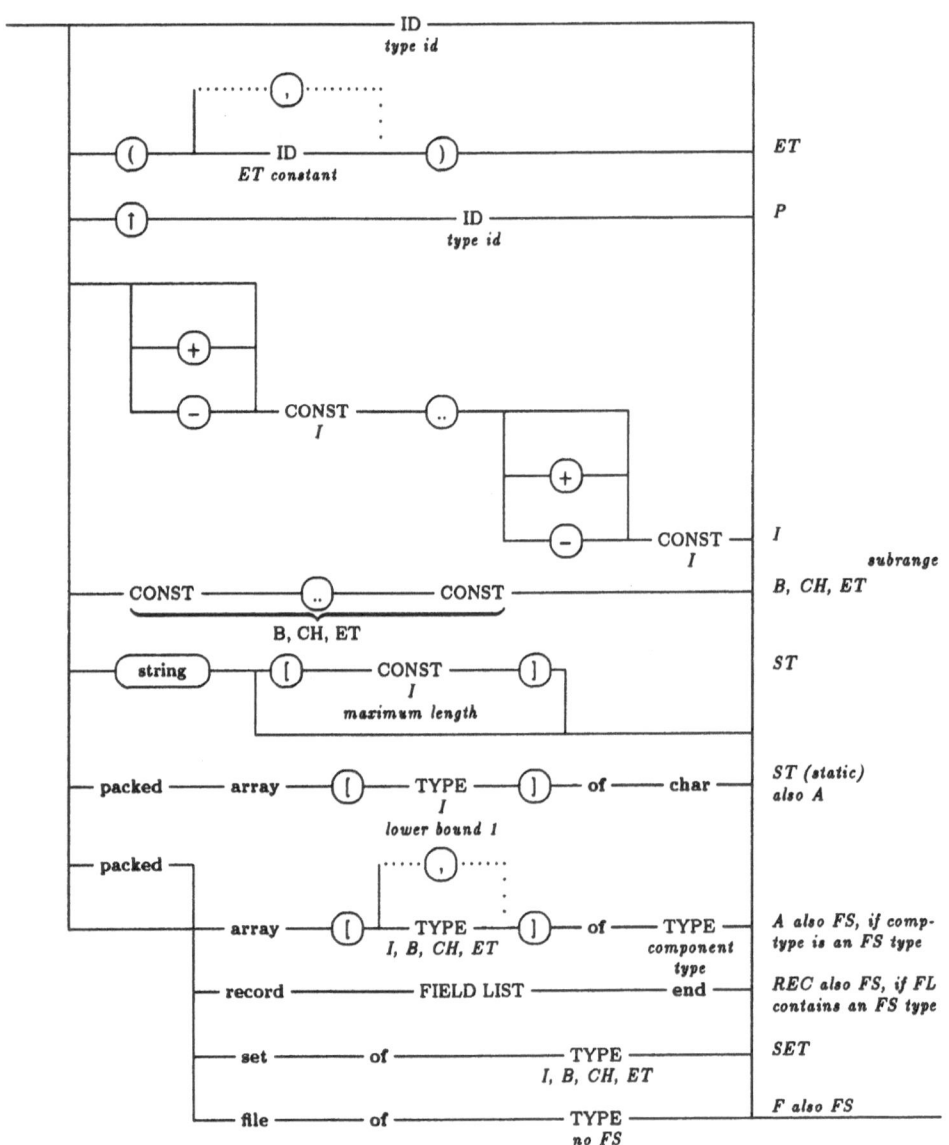

P11   FIELD LIST   (FLIST)

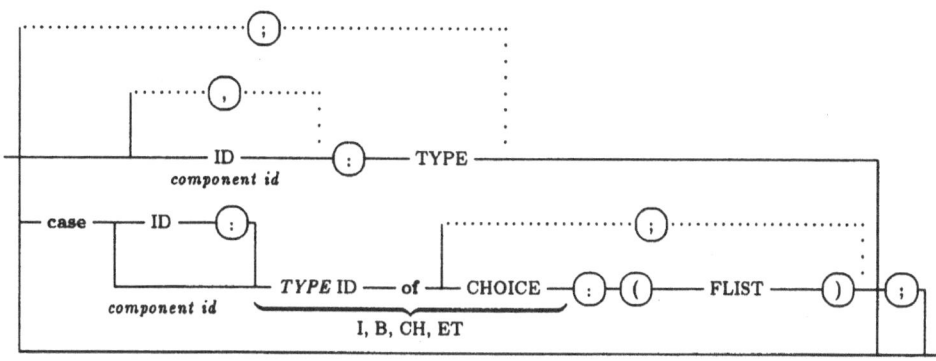

P12   VARIABLE DECLARATION

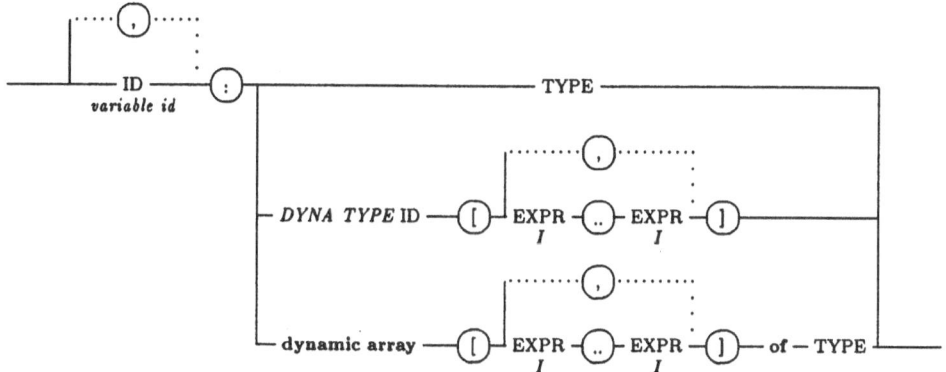

P13   CHOICE

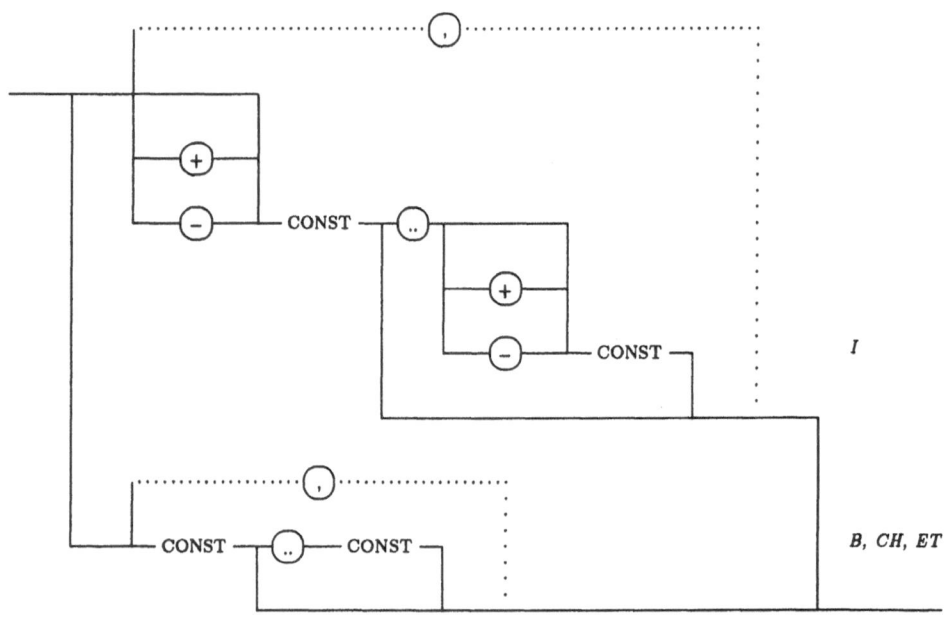

P14   PROCEDURE FUNCTION HEAD   (PFHEAD)

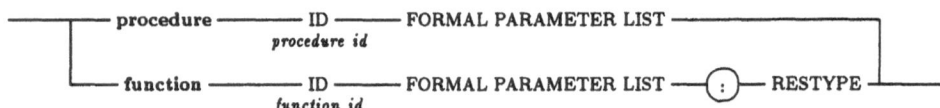

P15   PRIORITY DEFINITION

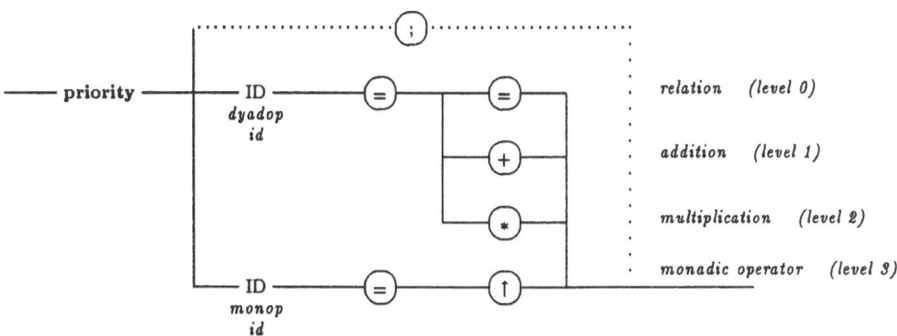

P16   OPERATOR HEAD

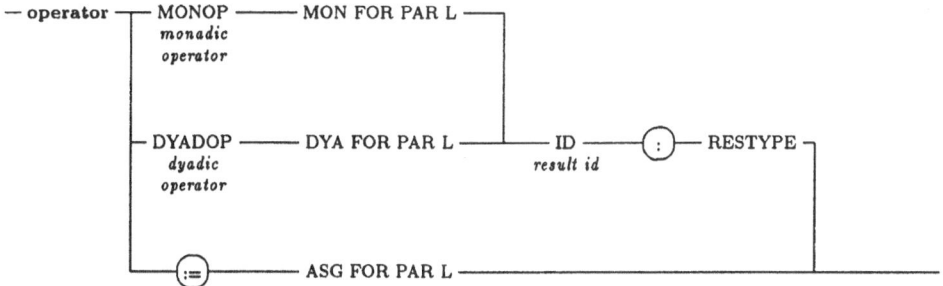

P17   MONADIC FORMAL PARAMETER LIST   (MON FOR PAR L)

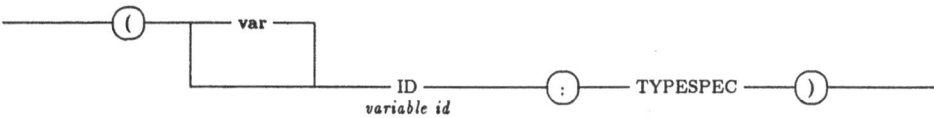

P18   DYADIC FORMAL PARAMETER LIST   (DYA FOR PAR L)

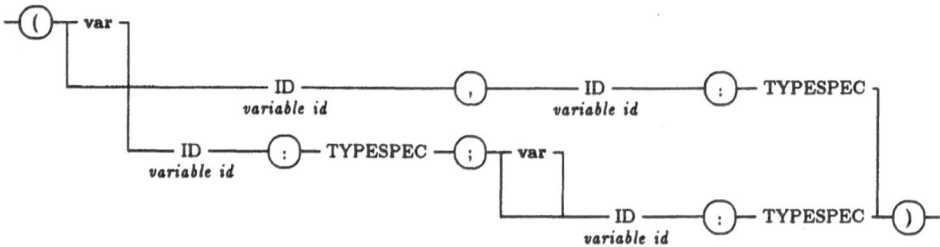

P19   ASSIGNMENT FORMAL PARAMETER LIST   (ASG FOR PAR L)

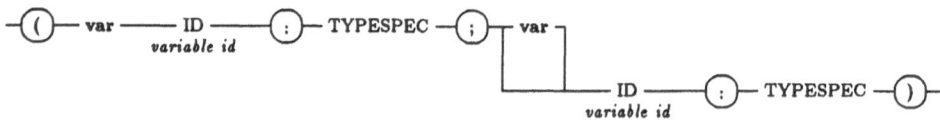

P20   RESULT TYPE   (RESTYPE)

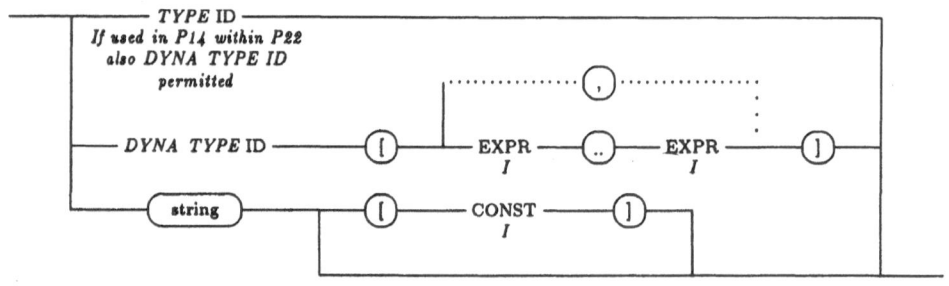

P21   BODY

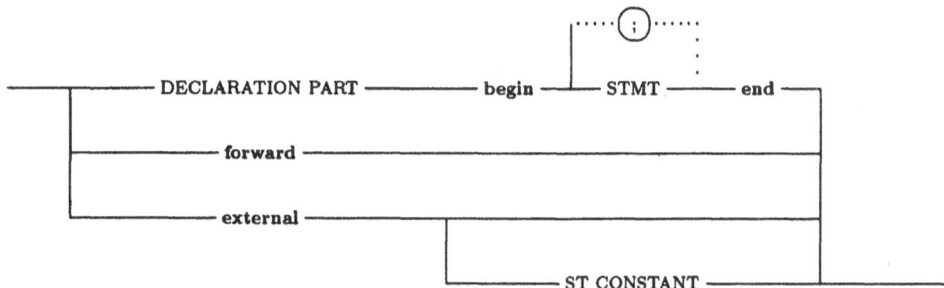

P22   FORMAL PARAMETER LIST

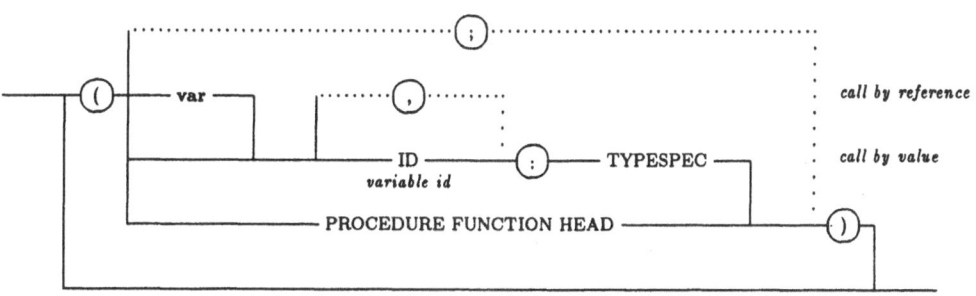

P23   TYPE SPECIFICATION   (TYPESPEC)

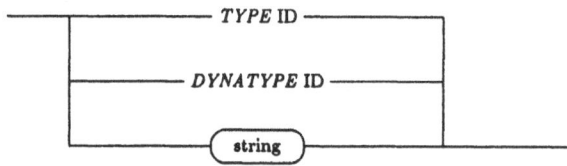

P24   STATEMENT   (STMT)

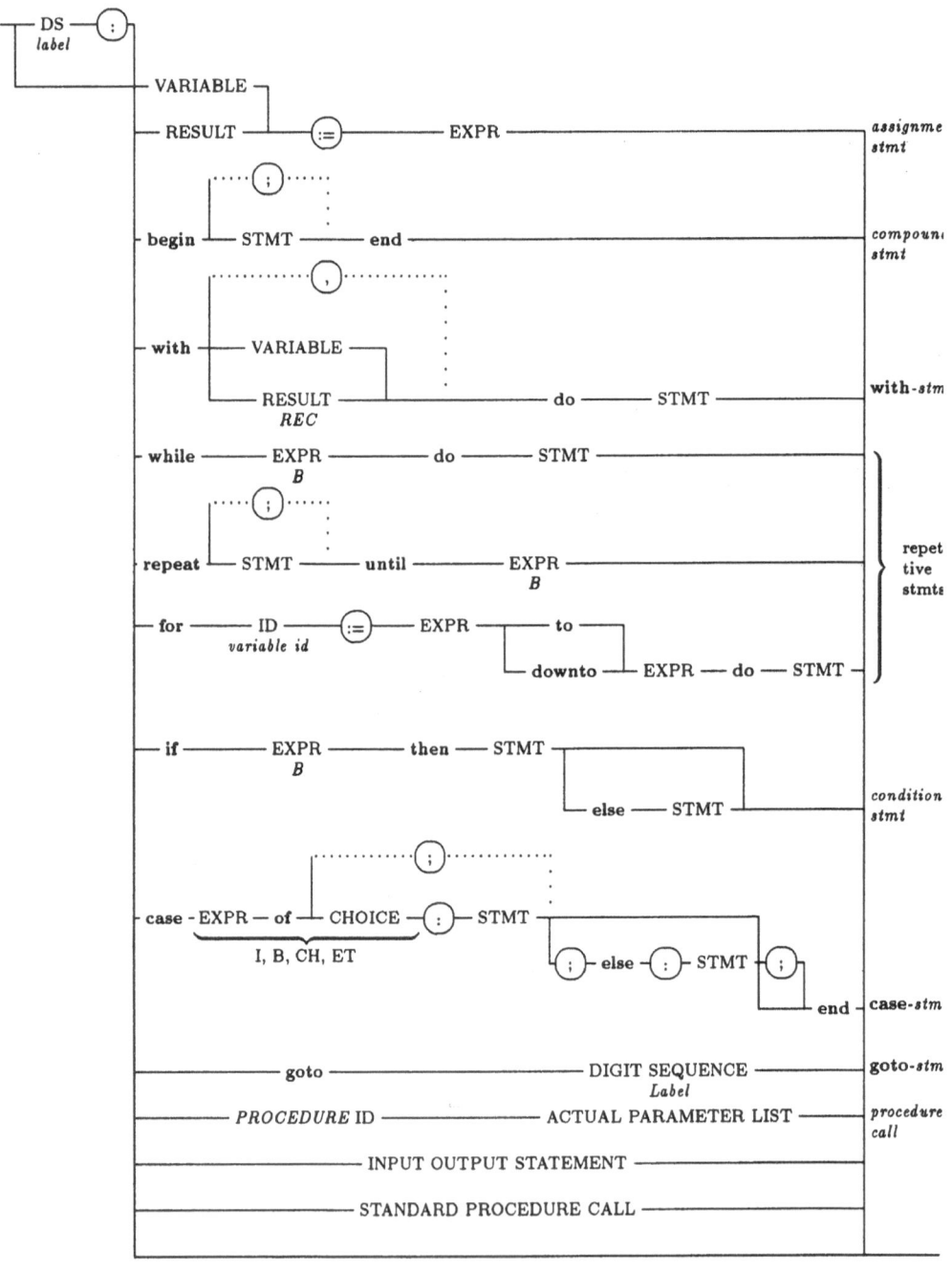

P25   RESULT

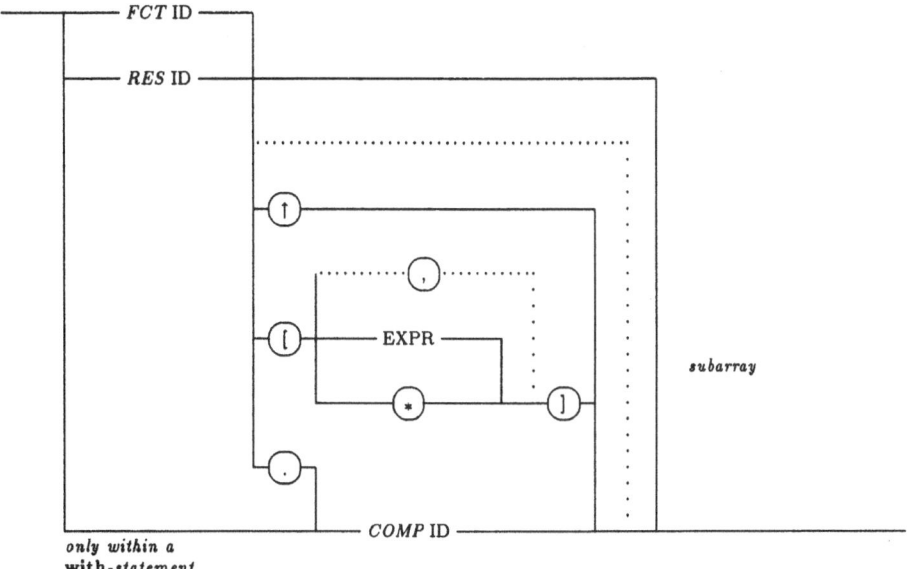

only within a
with-*statement*

Predefined PASCAL–XSC component identifiers

re,   im,   inf, sup

P26   STANDARD PROCEDURE CALL

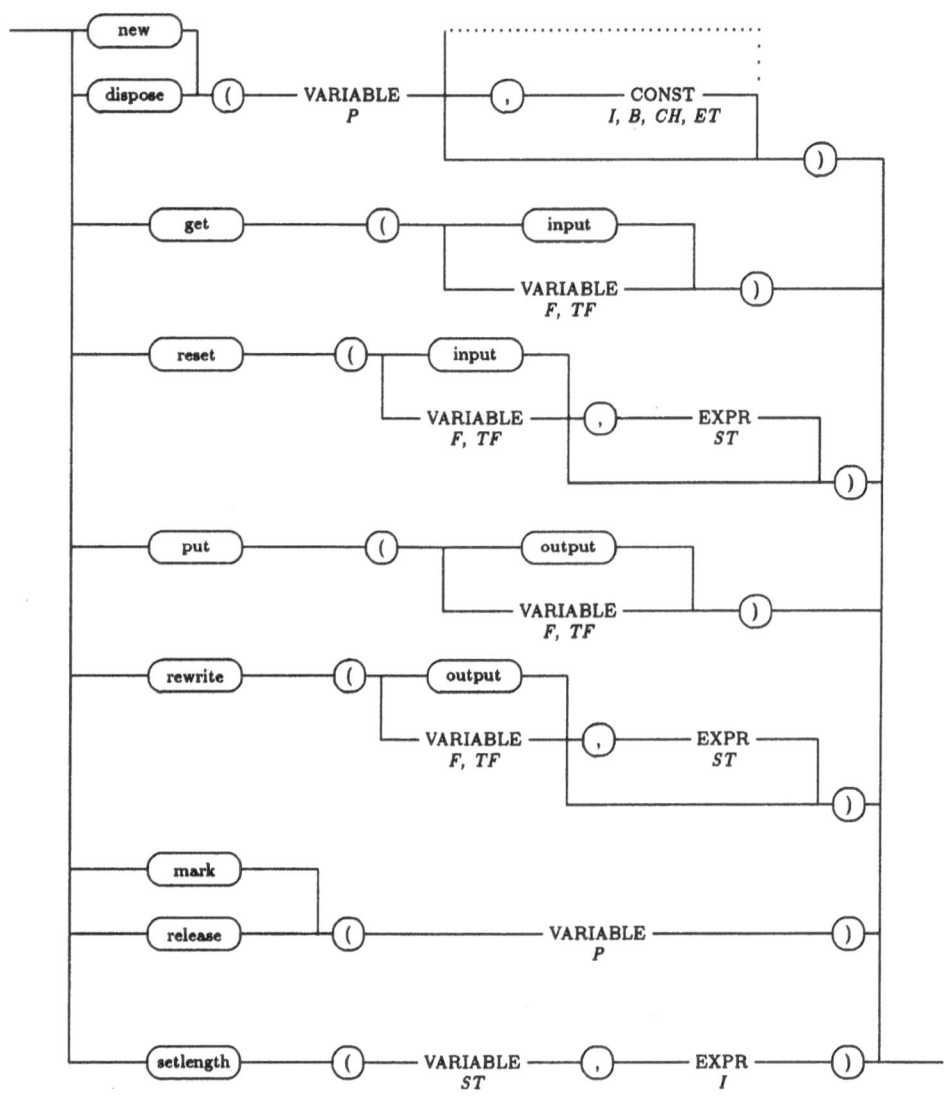

P27   INPUT OUTPUT STATEMENT   (IO STMT)

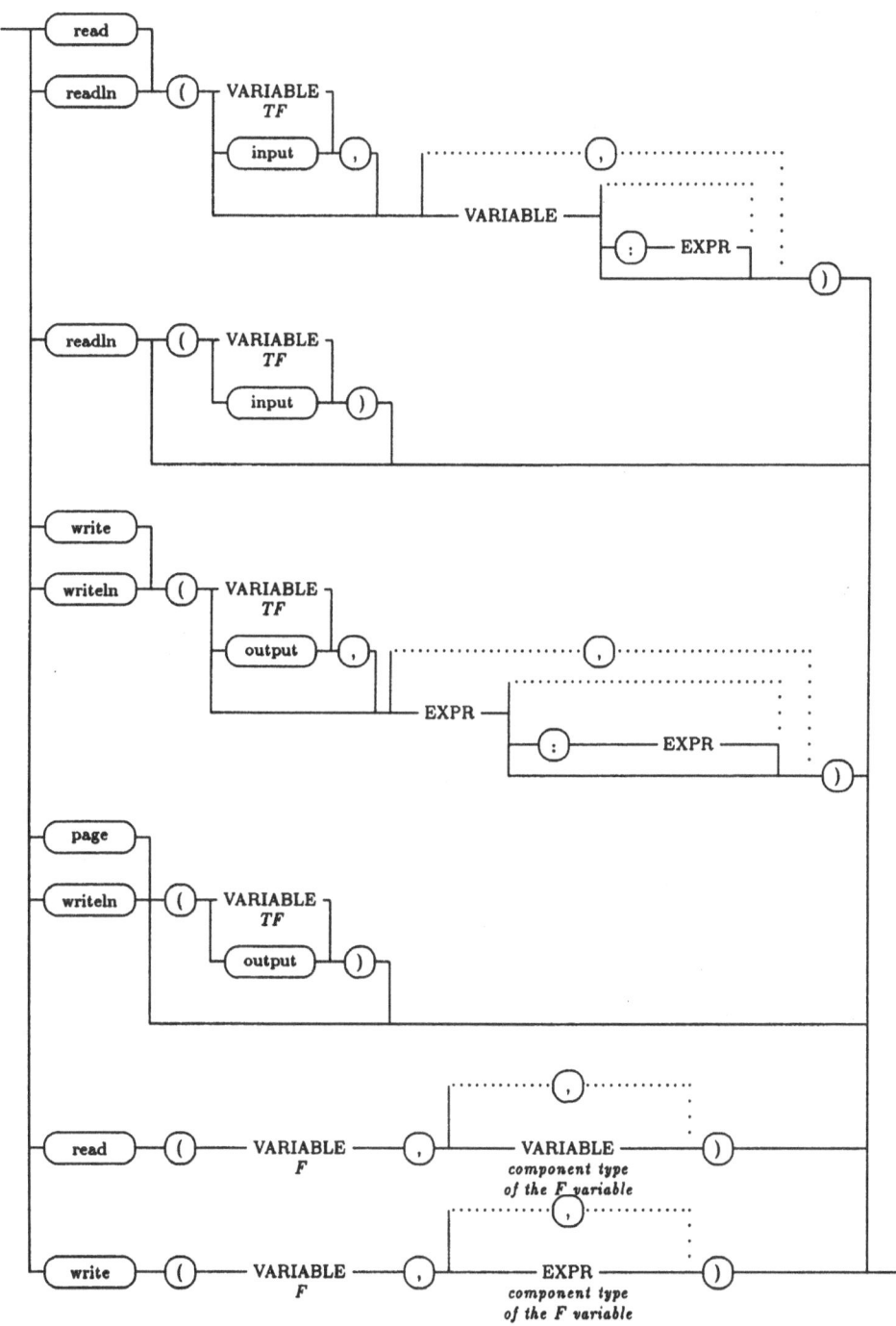

P28   EXPRESSION   (EXPR)

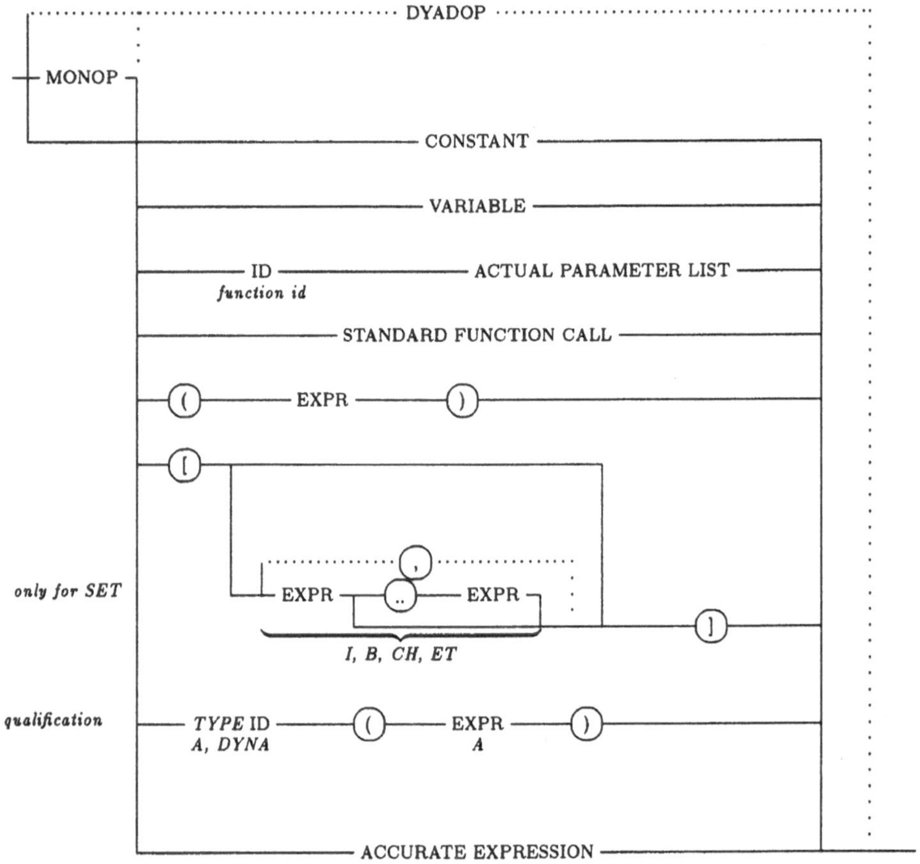

DYADOP

MONOP

CONSTANT

VARIABLE

ID — ACTUAL PARAMETER LIST
function id

STANDARD FUNCTION CALL

(— EXPR —)

[

only for SET   ,
                              EXPR — .. — EXPR

                              I, B, CH, ET

                                                                    ]

qualification — *TYPE* ID — (— EXPR —)
A, DYNA   *A*

ACCURATE EXPRESSION

P29   ACCURATE EXPRESSION

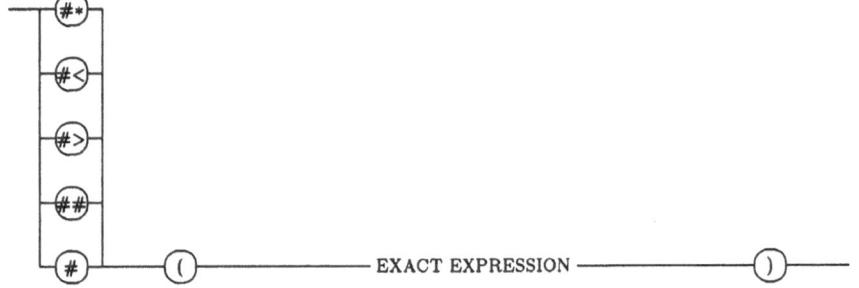

(#*)
(#<)
(#>)
(##)
(#) — (— EXACT EXPRESSION —)

P30   EXACT EXPRESSION   (EXA EXPR)

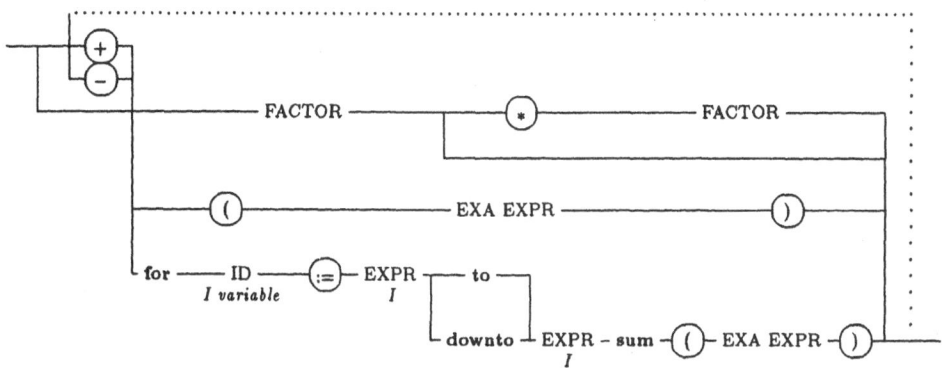

All Summands must have the same structure (scalar, vector, or matrix) and the same dimension.
No explicit accurate expressions are permitted in the I EXPR of the for-statement.

P31   FACTOR

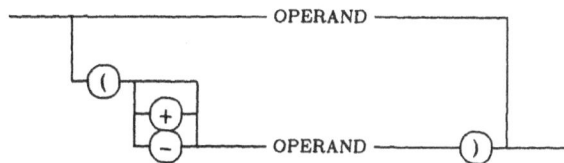

P32   OPERAND

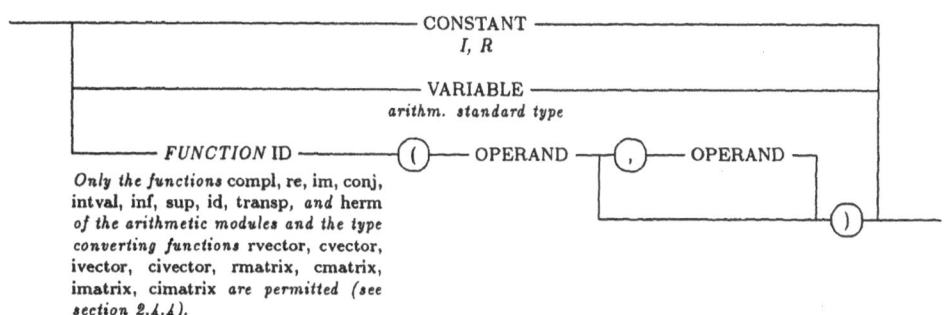

Only the functions compl, re, im, conj, intval, inf, sup, id, transp, *and* herm *of the arithmetic modules and the type converting functions* rvector, cvector, ivector, civector, rmatrix, cmatrix, imatrix, cimatrix *are permitted (see section 2.4.4).*

P33   DYADIC OPERATOR   (DYADOP)

```
─────────────── DYADOP ID ───────────┐
        ┌──────────── (=) ────────────┤ ┐
        ├──────────── (<>) ───────────┤ │
        ├──────────── (<=) ───────────┤ │
        ├──────────── (>=) ───────────┤ │
        ├──────────── (<) ────────────┤ ├ priority 0
        ├──────────── (>) ────────────┤ │
        ├──────────── in ─────────────┤ │
        ├──────────── (><) ───────────┤ ┘
        ├──────────── (+) ────────────┤ ┐
        ├──────────── (+<) ───────────┤ │
        ├──────────── (+>) ───────────┤ │
        ├──────────── (−) ────────────┤ │
        ├──────────── (−<) ───────────┤ ├ priority 1
        ├──────────── (−>) ───────────┤ │
        ├──────────── (+*) ───────────┤ │
        ├──────────── or ─────────────┤ ┘
        ├──────────── mod ────────────┤ ┐
        ├──────────── div ────────────┤ │
        ├──────────── (*) ────────────┤ │
        ├──────────── (*<) ───────────┤ │
        ├──────────── (*>) ───────────┤ │
        ├──────────── (/) ────────────┤ ├ priority 2
        ├──────────── (/<) ───────────┤ │
        ├──────────── (/>) ───────────┤ │
        ├──────────── (**) ───────────┤ │
        └──────────── and ────────────┘ ┘
```

P34   MONADIC OPERATOR   (MONOP)   priority 3

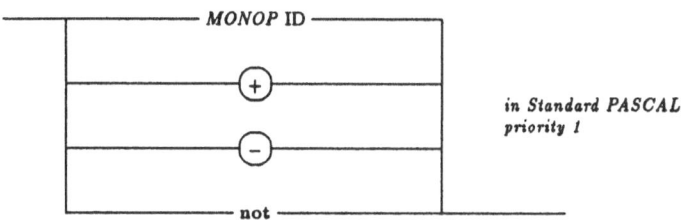

in Standard PASCAL priority 1

P35   ACTUAL PARAMETER LIST

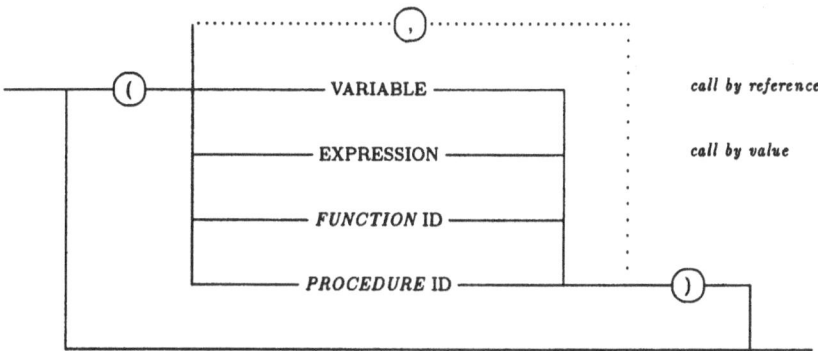

call by reference

call by value

P36   VARIABLE   (VAR)

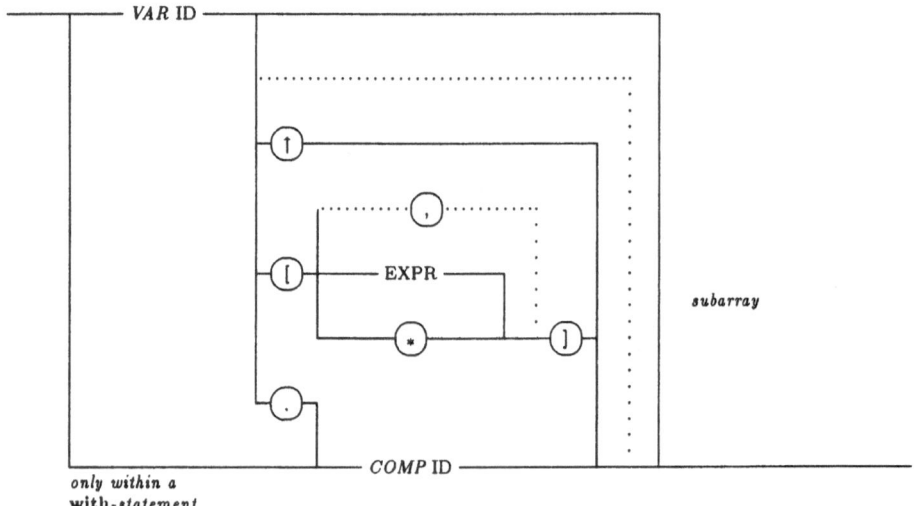

only within a
with-statement

Predefined PASCAL–XSC component identifiers

re,   im,   inf,   sup

P37   STANDARD FUNCTION CALL

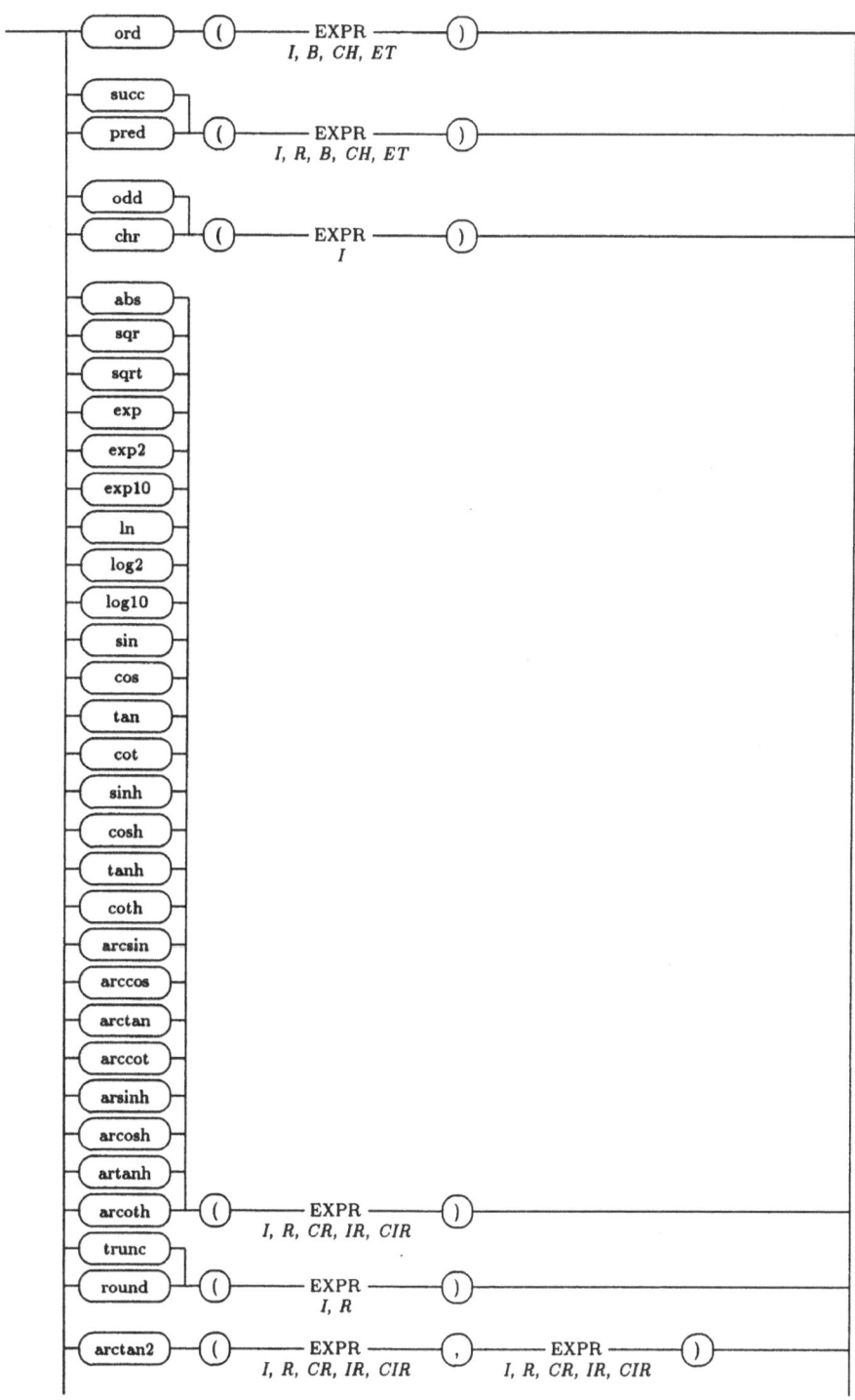

P37 STANDARD FUNCTION CALL (continued)

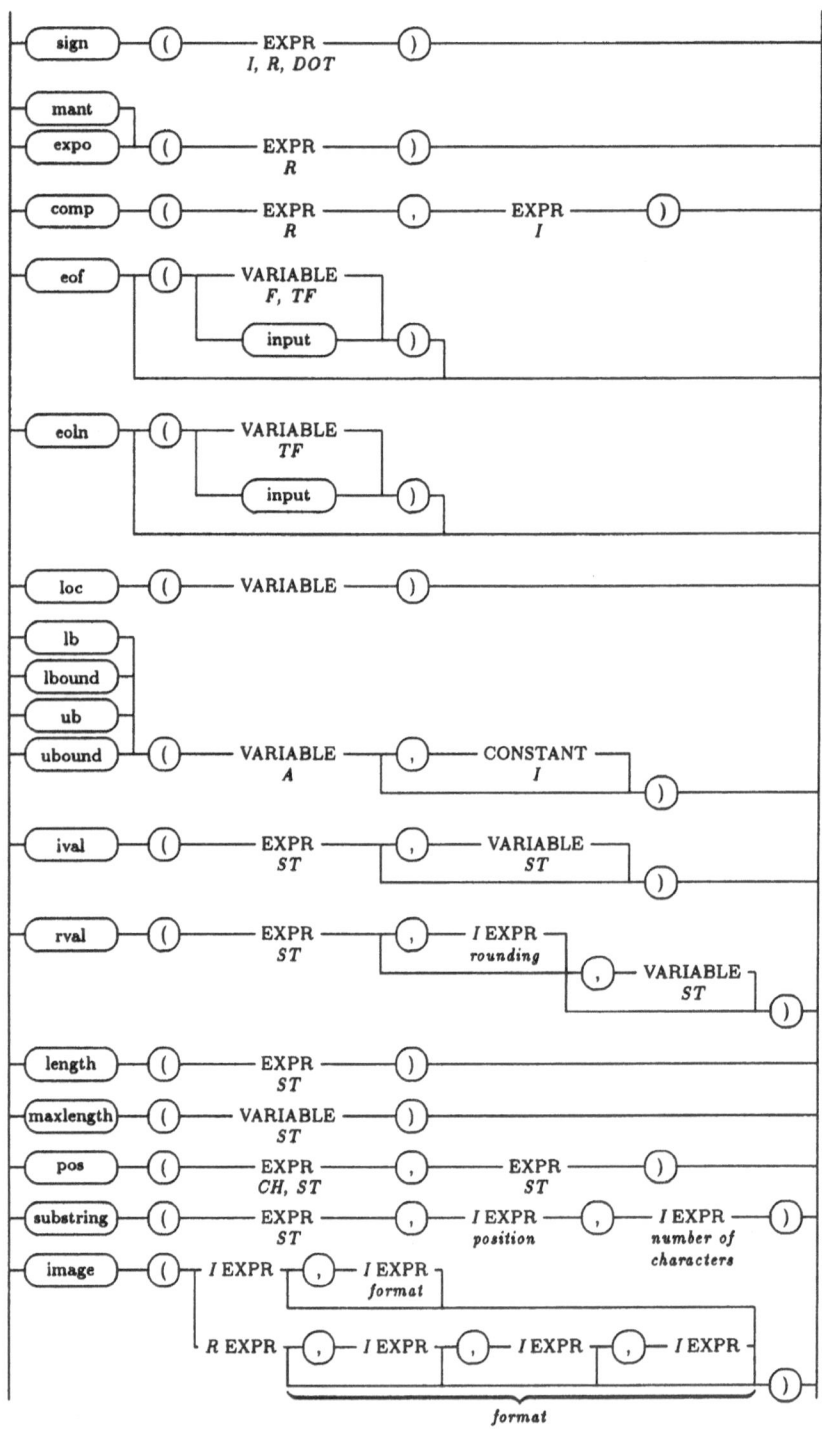

P37   STANDARD FUNCTION CALL   (continued; use of arithmetic modules assumed)

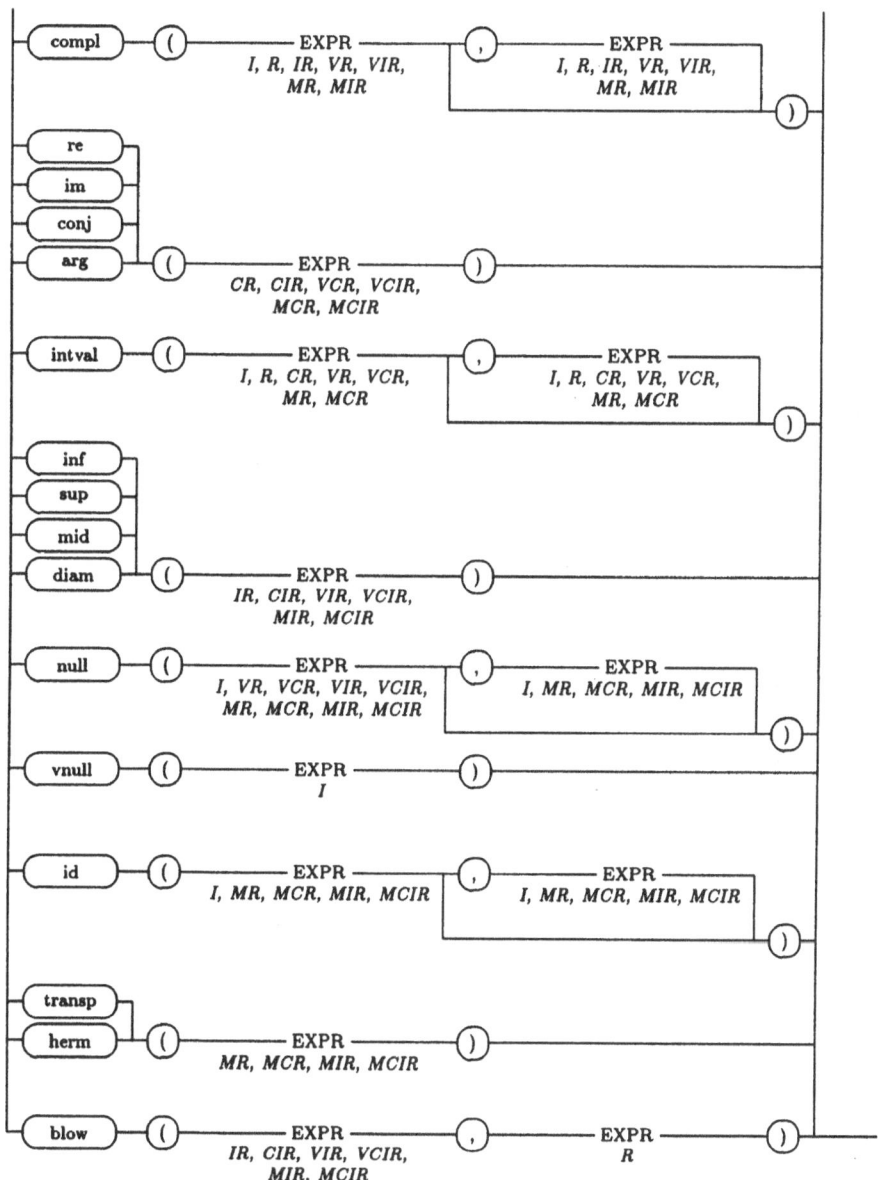

P38   IDENTIFIER   (ID)

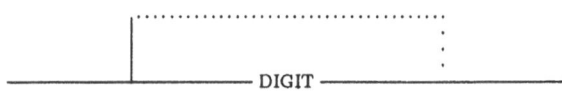

P39   DIGIT SEQUENCE   (DS)

P40   HEX DIGIT SEQUENCE   (HDS)

P41   CHARACTER

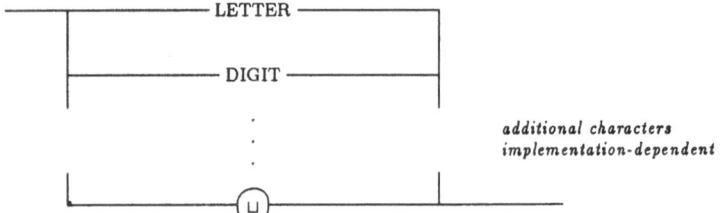

additional characters
implementation-dependent

P42   LETTER

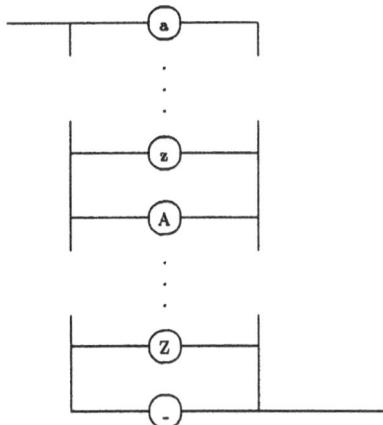

P43   DIGIT

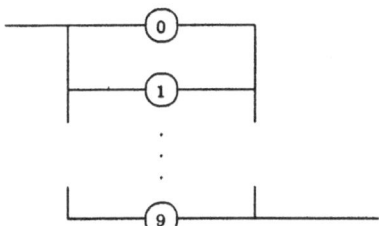

P44   HEX DIGIT

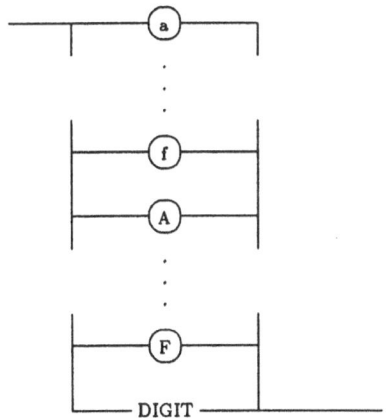

Appendix B

Indices and Lists

B.1 Syntax Diagrams

B.2   Reserved Words

and   array
begin
case   const
div   do   downto   dynamic
else   end   external
file   for   forward   function
global   goto
if   in
label
mod   module
nil   not
of   operator   or
packed   priority   procedure   program
record   repeat
set   sum
then   to   type
until   use
var
while   with

B.3   Predefined Identifiers

Subsequently, the predefined identifiers of the language core as well as those of the arithmetic modules are listed. The latter are marked by the use of *italicized letters*.

Constants	false maxint true
Types	boolean char cimatrix cinterval civector cmatrix complex cvector dotprecision imatrix integer interval ivector real rmatrix rvector string text
Variables	input output
Component Identifiers	im inf re sup
Functions	abs arccos arccot arcosh arcoth arcsin arctan arctan2 *arg* arsinh artanh *blow* chr comp *compl* *conj* cos cosh cot coth *diam* eof eoln exp exp2 exp10 expo *herm* *id* *im* image *inf* *intval* ival lb lbound length ln loc log2 log10 mant maxlength *mid* *null* odd ord pos pred *re* round rval sign sin sinh sqr sqrt substring succ *sup* tan tanh *transp* trunc ub ubound *vnull*

Procedures

dispose
get
mark
new
page   put
read   readln   release   reset   rewrite
write   writeln

B.4 Operators

The tables in this section list all of the predefined operators in the language core
and in the arithmetic modules.

B.4.1 Basic Operators

left operand \ right operand	integer	boolean	char	string	set
monadic	+, −	**not**			
integer	+, −, *, /, **div, mod,** ∨				**in**
boolean		**or, and,** =, <>, <=, >=			**in**
char			+ ∨	+ ∨ in	**in**
string			+ ∨	+ ∨ in	
set					+, −, *, =, <>, <=, >=
enumeration type					**in**

$$\vee \in \{=, <>, <, <=, >, >=\}$$

B.4.2 Arithmetic Operators

left operand \ right operand	integer real complex	interval cinterval	rvector cvector	ivector civector	rmatrix cmatrix	imatrix cimatrix
monadic [1]	$+,-$	$+,-$	$+,-$	$+,-$	$+,-$	$+,-$
integer real complex	$\circ, \circ<, \circ>,$ [2] $+*$	$+,-,*,/,$ $+*$	$*,*<,*>$	$*$	$*,*<,*>$	$*$
interval cinterval	$+,-,*,/,$ $+*$	$+,-,*,/,$ $+*,**$	$*$	$*$	$*$	$*$
rvector cvector	$*,*<,*>,$ $/,/<,/>$	$*,/$	$\circ, \circ<, \circ>,$ [3] $+*$	$+,-,*,$ [4] $+*$		
ivector civector	$*,/$	$*,/$	$+,-,*,$ [4] $+*$	$+,-,*,$ [4] $+*,**$		
rmatrix cmatrix	$*,*<,*>,$ $/,/<,/>$	$*,/$	$*,*<,*>$	$*$	$\circ, \circ<, \circ>,$ [3] $+*$	$+,-,*,$ [4] $+*$
imatrix cimatrix	$*,/$	$*,/$	$*$	$*$	$+,-,*,$ [4] $+*$	$+,-,*,$ [4] $+*,**$

[1]) The operators of this row are monadic (i.e. there is no left operand).

[2]) $\circ \in \{+,-,*,/\}$

[3]) $\circ \in \{+,-,*\}$, where $*$ denotes the scalar or matrix product.

[4]) $*$ denotes the scalar or matrix product.

$+*$: Interval hull (smallest interval enclosing both operands)

$**$: Interval intersection

B.4.3 Relational Operators for the Arithmetic Types

left operand \ right operand	integer real complex	interval cinterval	rvector cvector	ivector civector	rmatrix cmatrix	imatrix cimatrix
integer real complex	=,<>, <=,<, >=,>	in =,<>				
interval cinterval	=,<>	in,><,[1] =,<>, <=,<, >=,>				
rvector cvector			=,<>, <=,<, >=,>	in =,<>		
ivector civector			=,<>	in,><,[1] =,<>, <=,<, >=,>		
rmatrix cmatrix					=,<>, <=,<, >=,>	in =,<>
imatrix cimatrix					=,<>	in,><,[1] =,<>, <=,<, >=,>

[1]) The operators <= and < denote the "subset" relations; >= and > denote the "superset" relations.

>< : Test for disjointedness of intervals

in : Test for membership of a point in an interval or
Test for strict inclusion of an interval in the interior of an interval

B.4.4   Assignment Operators

The subsequent tables give a survey of all possible assignment statements which have been made possible by overloading of the operator := in the arithmetic modules.

Type of Left Side	Type of Right Side	Overloading Defined in
complex	integer real	module C_ARI
interval	integer real	module I_ARI
cinterval	integer real complex interval	module CI_ARI
rvector	integer real	module MV_ARI
cvector	integer real complex rvector	module MVC_ARI
ivector	integer real interval rvector	module MVI_ARI
civector	integer real complex interval cinterval rvector cvector ivector	module MVCI_ARI
rmatrix	integer real	module MV_ARI
cmatrix	integer real complex rmatrix	module MVC_ARI
imatrix	integer real interval rmatrix	module MVI_ARI

Type of Left Side	Type of Right Side	Overloading Defined in
cimatrix	integer real complex interval cinterval rmatrix cmatrix imatrix	module MVCLARI

B.5 Predefined Functions

In this section, we supply an alphabetical review of all predefined functions with their declaration (interface) and a short explanation of their purpose. For functions which are overloaded or newly defined in the arithmetical modules, the name of the defining module is listed. For the generic mathematical functions, consult the extra table at the end of this section. For details about the domain of definition or the range of the result, see the user manual of the compiler version you are using.

$\Big(\ \textbf{abs}\ \Big)$

    see extra table on page 318

$\Big(\ \textbf{arccos}\ \Big)$

    see extra table on page 318

$\Big(\ \textbf{arccot}\ \Big)$

    see extra table on page 318

$\Big(\ \textbf{arcosh}\ \Big)$

    see extra table on page 318

$\Big(\ \textbf{arcoth}\ \Big)$

    see extra table on page 318

$\Big(\ \textbf{arcsin}\ \Big)$

    see extra table on page 318

$\Big(\ \textbf{arctan}\ \Big)$

    see extra table on page 318

$\Big(\ \textbf{arctan2}\ \Big)$

    see extra table on page 318

$\Big(\ \textbf{arg}\ \Big)$

    function arg (c: complex) : real;

        Purpose: Delivers the argument (angle component) of the exponential representation of c.

        def./overl. in: C_ARI

function arg (c: cinterval) : interval;

    Purpose: Delivers the argument interval (angle component) of the exponential representation of c.

    def./overl. in: CI_ARI

arsinh

see extra table on page 318

artanh

see extra table on page 318

blow

function blow (x: Type1; eps: real) : Type1;

    Type1:   *interval, cinterval, ivector, civector, imatrix, cimatrix*

    Purpose: Delivers the epsilon inflation of the interval argument x (componentwise for array types). For x of type *interval*, *blow* is computed by

$$y \quad := (1 + eps) * x - eps * x;$$
$$blow := intval (\text{pred}(\inf(y)) , \text{succ}(\sup(y)));$$

    def./overl. in: I_ARI, CI_ARI, MVI_ARI, MVCI_ARI

chr

function chr (i: integer) : char;

    Purpose: Delivers the character with the ordinal number i. It is an error if no such value exists.

comp

function comp (m: real; e: integer) : real;

    Purpose: Composition of a mantissa m and an exponent e into a floating point value $m \cdot b^e$. It is an error if the values of b, e, and m do not lie in the implementation-dependent range.

conj

function conj (c: Type1) : Type1;

    Type1:   *complex, cinterval, cvector, civector, cmatrix, cimatrix*

    Purpose: Conjugation (for vector and matrix types in every component)

    def./overl. in: C_ARI, CI_ARI, MVC_ARI, MVCI_ARI

(**cos**)

see extra table on page 318

(**cosh**)

see extra table on page 318

(**cot**)

see extra table on page 318

(**coth**)

see extra table on page 318

(**diam**)

function diam (x: Type1) : ResType;

Type1:	*interval, cinterval, ivector, civector, imatrix, cimatrix*
ResType:	*real, rvector, rmatrix* according to the structure of Type1.
Purpose:	Delivers the diameter of x (for array types in every component).
def./overl. in:	LARI, CLARI, MVLARI, MVCLARI

(**eof**)

function eof (**var** f: Type1) : boolean;

Type1:	*text*, **file of** ...
Purpose:	Delivers *false* if the actual component of the file variable f is a defined component, otherwise *true*. It is an error if f is undefined.

function eof : boolean;

Purpose:	Corresponds to *eof (input)*.

(**eoln**)

function eoln (**var** f: text) : boolean;

Purpose:	Delivers *true* if the actual component of the file variable f contains the end-of-line character, otherwise *false*. It is an error if f is undefined.

function eoln : boolean;

Purpose:	Corresponds to *eoln (input)*.

(**exp**)

see extra table on page 318

(**exp2**)

see extra table on page 318

(**exp10**)

see extra table on page 318

(**expo**)

function expo (x: real) : integer;

Purpose: Delivers the exponent of x corresponding to the normal-
                  ized mantissa and the base.

(**herm**)

function herm (x: Type1) : Type1;

Type1:   *cmatrix, cimatrix*
Purpose: Delivers the Hermitean matrix.
def./overl. in: MVC_ARI, MVCI_ARI

(**id**)

function id (x: Type1) : rmatrix[lb(x)..ub(x),lb(x,2)..ub(x,2)];

Type1:   *rmatrix, cmatrix, imatrix, cimatrix*
Purpose: Delivers an identity matrix with the index range of x.
def./overl. in: MV_ARI, MVC_ARI, MVI_ARI, MVCI_ARI

function id (x, y: Type1) : rmatrix[lb(x)..ub(x),lb(y,2)..ub(y,2)];

Type1:   *rmatrix, cmatrix, imatrix, cimatrix*
Purpose: Delivers an identity matrix with the index ranges of the
                  product matrix x · y.
def./overl. in: MV_ARI, MVC_ARI, MVI_ARI, MVCI_ARI

function id (n: integer) : rmatrix[1..n,1..n];

Purpose: Delivers a n × n square identity matrix (n ≥ 1 assumed).
def./overl. in: MV_ARI

function id (n1, n2: integer) : rmatrix[1..n1,1..n2];

Purpose: Delivers a rectangular n1 × n2 identity matrix (n1, n2
                  ≥ 1 assumed).
def./overl. in: MV_ARI

(image)

function image (i: integer) : string;

Purpose: Converts the *integer* value *i* into a string with a current length according to the default output of *integer* values (possibly filled by leading blanks).

function image (i: integer; width: integer) : string;

Purpose: Converts the *integer* value *i* into a string with the length *width* (possibly filled with leading blanks).

function image (r: real) : string;

Purpose: Converts the *real* value *r* into a string with a current length according to the default output for *real* values (possibly filled by leading blanks).

function image (r: real; width: integer) : string;

Purpose: Converts the *real* value *r* into a string with the length *width* (possibly filled by leading blanks).

function image (r: real; width, fracs: integer) : string;

Purpose: Converts the *real* value *r* into a string with the length *width* (possibly filled by leading blanks) and *fracs* places after the decimal point.

function image (r: real; width, fracs, round: integer) : string;

Purpose: Converts the *real* value *r* into a string with the length *width* (possibly filled by leading blanks), *fracs* places after the decimal point, and rounded according to *round* (< 0 downwardly, $= 0$ to the nearest, > 0 upwardly).

(ival)

function ival (s: string) : integer;

Purpose: Converts the first part of the string *s*, which represents a numeric value according to the rules for *integer* constants, into an *integer* value. Leading blanks as well as trailing characters are neglected. It is an error if *s* does not satisfy the syntax of an *integer* constant.

function ival (s: string; **var** rest: string) : integer;

Purpose: Converts the first part of the string *s*, which represents a numeric value according to the rules for *integer* constants, into an *integer* value. Leading blanks are neglected, whereas trailing characters are passed back in the string *rest*. It is an error if *s* does not satisfy the syntax of an *integer* constant.

> **lb**

function lb (**var** a: Type1; i: integer) : ResType;

Type1:	Arbitrary array type
ResType:	Index type of Type1
Purpose:	Short form of *lbound*. Delivers the lower bound of the *i*-th index range of a. It is an error if *i* exceeds the number of dimensions.

function lb (**var** a: Type1) : ResType;

Type1:	Arbitrary array type
ResType:	Index type of Type1
Purpose:	Short form of *lbound*. Delivers the lower bound of the first index range of a.

> **lbound**

function lbound (**var** a: Type1; i: integer) : ResType;

Type1:	Arbitrary array type
ResType:	Index type of Type1
Purpose:	Delivers the lower bound of the *i*-th index range of a. It is an error if *i* exceeds the number of dimensions.

function lbound (**var** a: Type1) : ResType;

Type1:	Arbitrary array type
ResType:	Index type of Type1
Purpose:	Delivers the lower bound of the first index range of a.

> **length**

function length (s: string) : integer;

Purpose:	Delivers the current length of the *string* expression s.

> **ln**

see extra table on page 318

> **loc**

function loc (**var** x: Type1) : integer;

Type1:	Arbitrary type
Purpose:	Delivers the implementation-dependent memory address of the variable x.

$\boxed{\log 2}$

see extra table on page 318

$\boxed{\log 10}$

see extra table on page 318

$\boxed{\text{mant}}$

function mant (x: real) : real;

Purpose:	Delivers the normalized mantissa m (value range implementation-dependent) of x. It is an error if the values of x and m do not lie in the implementation-dependent range.

$\boxed{\text{maxlength}}$

function maxlength (**var** s: string) : integer;

Purpose:	Delivers the maximum length of the *string* variable s.

$\boxed{\text{mid}}$

function mid (x: Type1) : ResType;

Type1:	*interval, cinterval, ivector, civector, imatrix, cimatrix*
ResType:	Type of the lower bound (*inf*) or of the upper bound (*sup*) of Type1.
Purpose:	Delivers the midpoint of x (in each component for array types).
def./overl. in:	I_ARI, CI_ARI, MVI_ARI, MVCI_ARI

$\boxed{\text{null}}$

function null (x: Type1) : rvector[lb(x)..ub(x)];

Type1:	*rvector, cvector, ivector, civector*
Purpose:	Delivers a zero vector with the index range of x.
def./overl. in:	MV_ARI, MVC_ARI, MVI_ARI, MVCI_ARI

function null (x: Type2) : rmatrix[lb(x)..ub(x),lb(x,2)..ub(x,2)];

Type2:	*rmatrix, cmatrix, imatrix, cimatrix*
Purpose:	Delivers a zero matrix with the index ranges of x.
def./overl. in:	MV_ARI, MVC_ARI, MVI_ARI, MVCI_ARI

function null (x, y: Type2) : rmatrix[lb(x)..ub(x),lb(y,2)..ub(y,2)];

      Type2:   *rmatrix, cmatrix, imatrix, cimatrix*

      Purpose: Delivers a zero matrix with the index ranges of the product matrix $x \cdot y$.

      def./overl. in: MV_ARI, MVC_ARI, MVI_ARI, MVCI_ARI

function null (n: integer) : rmatrix[1..n,1..n];

      Purpose: Delivers a $n \times n$ square zero matrix ($n \geq 1$ assumed).

      def./overl. in: MV_ARI

function null (n1, n2: integer) : rmatrix[1..n1,1..n2];

      Purpose: Delivers an $n1 \times n2$ zero matrix ($n1, n2 \geq 1$ assumed).

      def./overl. in: MV_ARI

---

odd

function odd (i: integer) : boolean;

      Purpose: Delivers *true* if i is an odd number, otherwise *false*.

---

ord

function ord (x: Type1) : integer;

      Type1:   *integer, boolean, char, enumeration type, pointer type*

      Purpose: Delivers the ordinal number of x or the value of the pointer, if x is of pointer type.

---

pos

function pos (s1, s2: string) : integer;

      Purpose: Delivers the position of the first occurrence of $s1$ in $s2$.

---

pred

function pred (x: Type1) : Type1;

      Type1:   *integer, real, boolean, char, enumeration type*

      Purpose: Delivers the predecessor of x. It is an error if no predecessor exists.

round

function round (x: Type1) : integer;

Type1:	*integer, real*
Purpose:	Rounding to the nearest *integer* number. The result satisfies

$$\text{round } (x) = \text{sign } (x) * \text{trunc } (\text{abs } (x) + 0.5).$$

It is an error if no such *integer* value exists.

rval

function rval (s: string) : real;

Purpose:	Converts the first part of the string *s*, which represents a numeric value according to the rules of *real* constants, into a *real* value. Leading blanks as well as trailing characters are neglected. It is an error if *s* does not satisfy the syntax of an *real* constant.

function rval (s: string; **var** rest: string) : real;

Purpose:	Converts the first part of the string *s*, which represents a numeric value according to the rules of *real* constants, into a *real* value. Leading blanks are neglected, whereas trailing characters are passed back in the string *rest*. It is an error if *s* does not satisfy the syntax of an *real* constant.

function rval (s: string; round: integer) : real;

Purpose:	Converts the first part of the string *s*, which represents a numeric value according to the rules of *real* constants, into a *real* value rounded according to *round* (< 0 downwardly, $= 0$ to the nearest, > 0 upwardly). Leading blanks as well as trailing characters are neglected. It is an error if *s* does not satisfy the syntax of an *real* constant.

function rval (s: string; round: integer; **var** rest: string) : real;

Purpose:	Converts the first part of the string *s*, which represents a numeric value according to the rules of *real* constants, into a *real* value rounded according to *round* (< 0 downwardly, $= 0$ to the nearest, > 0 upwardly). Leading blanks are neglected, whereas trailing characters are passed back in the string *rest*. It is an error if *s* does not satisfy the syntax of an *real* constant.

(**sign**)

> **function** sign (x: Type1) : integer;
>
> > Type1:   *integer, real, dotprecision*
> > Purpose: Delivers the sign of x (−1 for x < 0, 1 for x > 0, 0 for x = 0).

(**sin**)

> see extra table on page 318

(**sinh**)

> see extra table on page 318

(**sqr**)

> see extra table on page 318

(**sqrt**)

> see extra table on page 318

(**substring**)

> **function** substring (s: string; pos, number: integer) : string;
>
> > Purpose: Returns a substring of s containing *number* characters starting from position *pos*. If *pos* is larger than the current length of s, an empty string is returned. If s is shorter than *pos* + *number* characters, a shorter string is returned. For *pos* < 1, *pos* is set to 1.

(**succ**)

> **function** succ (x: Type1) : Type1;
>
> > Type1:   *integer, real, boolean, char, enumeration type*
> > Purpose: Delivers the successor of x. It is an error if no successor exists.

(**tan**)

> see extra table on page 318

(**tanh**)

> see extra table on page 318

(**transp**)

function transp (x: Type1) : Type1;

Type1:	*rmatrix, cmatrix, imatrix, cimatrix*
Purpose:	Delivers the transposed matrix of x.
def./overl. in:	MV_ARI, MVC_ARI, MVI_ARI, MVCI_ARI

(**trunc**)

function trunc (x: Type1) : integer;

Type1:	*integer, real*
Purpose:	Rounding to an *integer* number by truncation of the fractional portion of x. It is an error if no such *integer* value exists.

(**ub**)

function ub (**var** a: Type1; i: integer) : ResType;

Type1:	Arbitrary array type
ResType:	Index type of Type1
Purpose:	Short form of *ubound*. Delivers the upper bound of the *i*-th index range of a. It is an error if *i* exceeds the number of dimensions.

function ub (**var** a: Type1) : ResType;

Type1:	Arbitrary array type
ResType:	Index type of Type1
Purpose:	Short form of *ubound*. Delivers the upper bound of the first index range of a.

(**ubound**)

function ubound (**var** a: Type1; i: integer) : ResType;

Type1:	Arbitrary array Type
ResType:	Index type of Type1
Purpose:	Delivers the upper bound of the *i*-th index range of a. It is an error if *i* exceeds the number of dimensions.

function ubound (**var** a: Type1) : ResType;

Type1:	Arbitrary array type
ResType:	Index type of Type1
Purpose:	Delivers the upper bound of the first index range of a.

 (**vnull**)

function vnull (n: integer) : rvector[1..n];

Purpose: Delivers a zero vector with n components ($n \geq 1$ assumed).

def./overl. in: MV_ARI

The Predefined Mathematical Functions

	Function	Generic Name
1	Absolute Value	abs
2	Arc Cosine	arccos
3	Arc Cotangent	arccot
4	Inverse Hyperbolic Cosine	arcosh
5	Inverse Hyperbolic Cotangent	arcoth
6	Arc Sine	arcsin
7	Arc Tangent	arctan
8	Inverse Hyperbolic Sine	arsinh
9	Inverse Hyperbolic Tangent	artanh
10	Cosine	cos
11	Cotangent	cot
12	Hyperbolic Cosine	cosh
13	Hyperbolic Cotangent	coth
14	Exponential Function	exp
15	Power Function (Base 2)	exp2
16	Power Function (Base 10)	exp10
17	Natural Logarithm (Base e)	ln
18	Logarithm (Base 2)	log2
19	Logarithm (Base 10)	log10
20	Sine	sin
21	Hyperbolic Sine	sinh

	Function	Generic Name
22	Square	sqr
23	Square Root	sqrt
24	Tangent	tan
25	Hyperbolic Tangent	tanh

The argument type for each of these functions may be any of the types *integer*, *real*, *complex*, *interval*, and *cinterval*, i. e. the functions are defined not only for the types *integer* and *real*. They are also provided for the types *complex*, *interval*, and *cinterval* in the arithmetic modules C_ARI, I_ARI, and CI_ARI.

We do not explain the interfaces and the formal declarations, because all of these functions are defined with only one formal parameter. Normally, the result type is the same as the type of the argument. For *integer* arguments, this only holds for the functions *abs* and *sqr*. All other functions return *real* values for *integer* arguments.

In addition to the standard functions listed in the table above, the function

    arctan2 (x1,x2)

is available for two arguments *x1*, *x2* of type *real* or *interval*. The result of *arctan2* *(x1, x2)* is

    arctan (x1/x2) .

B.6   Transfer Functions

In this section, we give an alphabetical review of the transfer functions for type conversion between the arithmetic types. Beneath their declaration (interface) and a short explanation of their purpose, we list the name of the module, in which the functions are defined or overloaded.

$$\boxed{\text{compl}}$$

function compl (x1: Type1; x2: Type2) : Type3;

Type1:	*real, interval, rvector, ivector, rmatrix, imatrix*
Type2:	*real, interval, rvector, ivector, rmatrix, imatrix* with the corresponding structure (scalar, vector, matrix) of Type1
Type3:	Corresponding complex type of Type1 or Type2 (*complex, cinterval, cvector, civector, cmatrix, cimatrix*)
Purpose:	Composition of the arguments *x1* and *x2* (real and imaginary parts) into the corresponding complex type (componentwise for vector and matrix types).
def./overl. in:	C_ARI, CI_ARI, MVC_ARI, MVCI_ARI

function compl (x: Type1) : Type2;

Type1:	*real, interval, rvector, ivector, rmatrix, imatrix*
Type2:	Corresponding complex type of Type1 (*complex, cinterval, cvector, civector, cmatrix, cimatrix*)
Purpose:	Composition of the argument *x* (real part) and imaginary part 0 to the corresponding complex type (componentwise for vector and matrix types).
def./overl. in:	C_ARI, CI_ARI, MVC_ARI, MVCI_ARI

$$\boxed{\text{im}}$$

function im (c: Type1) : Type2;

Type1:	*complex, cinterval, cvector, civector, cmatrix, cimatrix*
Type2:	Corresponding real or interval type of Type1 (*real, interval, rvector, ivector, rmatrix, imatrix*)
Purpose:	Delivers the imaginary part of the argument (componentwise for vector and matrix types).
def./overl. in:	C_ARI, CI_ARI, MVC_ARI, MVCI_ARI

(**inf**)

function inf (i: Type1) : Type2;

Type1:	*interval, cinterval, ivector, civector, imatrix, cimatrix*
Type2:	Corresponding real or complex type of Type1 (*real, complex, rvector, cvector, rmatrix, cmatrix*)
Purpose:	Delivers the lower bound of the interval argument (componentwise for vector and matrix types).
def./overl. in:	I_ARI, CI_ARI, MVI_ARI, MVCI_ARI

(**intval**)

function intval (x1: Type1; x2: Type2) : Type3;

Type1:	*real, complex, rvector, cvector, rmatrix, cmatrix*
Type2:	*real, complex, rvector, cvector, rmatrix, cmatrix* with the structure (scalar, vector, matrix) of Type1.
Type3:	Corresponding interval type of Type1 or Type2 (*interval, cinterval, ivector, civector, imatrix, cimatrix*)
Purpose:	Composition of the arguments *x1* and *x2* (lower and upper bound) to the corresponding interval type (componentwise for vector and matrix types). It is an error if $x1 > x2$.
def./overl. in:	I_ARI, CI_ARI, MVI_ARI, MVCI_ARI

function intval (x: Type1) : Type2;

Type1:	*real, complex, rvector, cvector, rmatrix, cmatrix*
Type2:	Corresponding interval type of Type1 (*interval, cinterval, ivector, civector, imatrix, cimatrix*)
Purpose:	Converting of the argument *x* into an interval with lower and upper bound equal to *x* (componentwise for vector and matrix types).
def./overl. in:	I_ARI, CI_ARI, MVI_ARI, MVCI_ARI

(**re**)

function re (c: Type1) : Type2;

Type1:	*complex, cinterval, cvector, civector, cmatrix, cimatrix*
Type2:	Corresponding real or interval type of Type1 (*real, interval, rvector, ivector, rmatrix, imatrix*)
Purpose:	Delivers the real part of the argument (componentwise for vector and matrix types).
def./overl. in:	C_ARI, CI_ARI, MVC_ARI, MVCI_ARI

```
(  sup  )
```

function sup (i: Type1) : Type2;

Type1:	*interval, cinterval, ivector, civector, imatrix, cimatrix*
Type2:	Corresponding real or complex type of Type1 (*real, complex, rvector, cvector, rmatrix, cmatrix*)
Purpose:	Delivers the upper bound of the interval argument (componentwise for vector and matrix types).
def./overl. in:	I_ARI, CI_ARI, MVI_ARI, MVCI_ARI

B.7   Predefined Procedures

In this section, we give an alphabetical review of the predefined procedures (including the input/output procedures) with their declaration part (interface) and a short explanation of their purpose. For the functions which are overloaded or newly defined in the arithmetic modules, we list the name of the defining module.

(**dispose**)

> **procedure** dispose (**var** p: Type1);
>
> | Type1: | Arbitrary pointer type |
> |---|---|
> | Purpose: | Release of the storage space of an element referenced by the pointer p. It is an error if $p =$ **nil**. The procedure *dispose* may not be used in conjunction with *release*. |
>
> **procedure** dispose (**var** p: Type1; c1,c2,...,cn: Type2);
>
> | Type1: | Arbitary pointer type |
> |---|---|
> | Type2: | *integer*, *boolean*, *char*, enumeration type |
> | Purpose: | Release of the storage space of an element referenced by the pointer p. The constants $c1$, ..., cn enable the access of special variants (for variant records). It is an error if $p =$ **nil**. The procedure *dispose* may not be used in conjunction with *release*. |

(**get**)

> **procedure** get (**var** f: Type1);
>
> | Type1: | *text*, **file of** ... |
> |---|---|
> | Purpose: | The next component of the actual component of the file variable f becomes the new actual component. The value of the actual component is assigned to the buffer variable $f\uparrow$. It is an error if f is undefined or if f is not in reading mode. |

(**mark**)

> **procedure** mark (**var** p: Type1);
>
> | Type1: | Arbitrary pointer type |
> |---|---|
> | Purpose: | Marks the *heap* to enable a *release* later. |

(**new**)

> **procedure** new (**var** p: Type1);
>
> | Type1: | Arbitrary pointer type |
> |---|---|
> | Purpose: | Creation of a new element referenced by the pointer p. |

procedure new (**var** p: Type1; c1,c2,...,cn: Type2);

Type1:	Arbitrary pointer type
Type2:	*integer*, *boolean*, *char*, enumeration type
Purpose:	Creation of a new element, referenced by the pointer p. The constants *c1*, ..., *cn* enable the access to special varaints (for variant records).

⬭ **page** ⬭

procedure page (**var** f: text);

Purpose:	Beginning of a new page on the output file *f*. It is an error if *f* is undefined or if *f* is not in writing mode.

procedure page;

Purpose:	Corresponds to *page* (*output*).

⬭ **put** ⬭

procedure put (**var** f: Type1);

Type1:	*text*, **file of** ...
Purpose:	The value of the buffer variable *f↑* is assigned to the actual component of *f*. The next component of the actual component of the file variable *f* becomes actual component. It is an error if *f* is undefined or if *f* is not in writing mode.

⬭ **read** ⬭

procedure read (**var** f: Type1; **var** x: Type2);

Type1:	*text*, **file of** ...
Type2:	*integer, char, string, real, complex, interval, cinterval, rvector, cvector, ivector, civector, rmatrix, cmatrix, imatrix, cimatrix*
Purpose:	Input of one or several variables of type *Type2* from file *f* (depending on *Type2*, format specifications are permitted seperated by colons). It is an error if *f* is undefined or if *f* is not in reading mode.
def./overl. in:	C_ARI, L_ARI, CI_ARI, MV_ARI, MVC_ARI, MVI_ARI and MVCI_ARI

procedure read (**var** x: Type2);

Type2:	*integer, char, string, real, complex, interval, cinterval, rvector, cvector, ivector, civector, rmatrix, cmatrix, imatrix, cimatrix*
Purpose:	Corresponds to *read*(*input*, x).

(**readln**)

procedure readln (**var** f: text);

    Purpose: Terminate an input line by reading the end-of-line char-
                           acter. It is an error if f is undefined or if f is not in
                           reading mode.

procedure readln;

    Purpose: Corresponds to *readln(input)*.

procedure readln (**var** f: text; **var** x: Type2);

    Type2:   *integer, char, string, real, complex, interval, cinter-*
                           val, rvector, cvector, ivector, civector, rmatrix, cmatrix,
                           imatrix, cimatrix

    Purpose: Corresponds to *read(f, x)* followed by *readln(f)*.

procedure readln (**var** x: Type2);

    Type2:   *integer, char, string, real, complex, interval, cinter-*
                           val, rvector, cvector, ivector, civector, rmatrix, cmatrix,
                           imatrix, cimatrix

    Purpose: Corresponds to *read(x)* followed by *readln*.

(**release**)

procedure release (**var** p: Type1);

    Type1:   Arbitrary pointer type
    Purpose: Refreshes the old state of *heap* marked by *mark*. All
                           variables created since the call of *mark* are released. The
                           pointer p must have the same value as the pointer used
                           for the most recent call of the procedure *mark*. The
                           procedure *mark* may not be used in conjunction with
                           dispose.

(**reset**)

procedure reset (**var** f: Type1);

    Type1:   *text*, **file of** ...
    Purpose: The file f is initialized for reading (input).

procedure reset (**var** f: Type1; s: string);

    Type1:   *text*, **file of** ...

    Purpose: The file *f* is initialized for reading (input). The physical file with external name *s* is associated with the internal file *f*.

(**rewrite**)

procedure rewrite (**var** f: Type1);

    Type1:   *text*, **file of** ...

    Purpose: The file *f* is initialized for writing (output).

procedure rewrite (**var** f: Type1; s: string);

    Type1:   *text*, **file of** ...

    Purpose: The file *f* is initialized for writing (output). The physical file with external name *s* is associated with the internal file *f*.

(**setlength**)

procedure setlength (**var** s: Type1; i: Type2);

    Type1:   *string[m]* or *string*

    Type2:   *0..m* or *0..M*

    Purpose: The actual length of the string variable *s* is set to *i* with $0 \leq i \leq m$ or $0 \leq i \leq M$, respectively, where M is the implementation-defined maximum length of strings. It is an error if *i* exceeds the maximum string length.

(**write**)

procedure write (**var** f: Type1; x: Type2);

    Type1:   *text*, **file of** ...

    Type2:   *integer, boolean, char, string, real, complex, interval, cinterval, rvector, cvector, ivector, civector, rmatrix, cmatrix, imatrix, cimatrix*

    Purpose: Output of one or several expressions of type *Type2* into the file *f* (depending on *Type2*, format specification are permitted seperated by colons). It is an error if *f* is undefined or if *f* is not in writing mode.

    def./overl. in: C_ARI,   I_ARI,   CI_ARI,   MV_ARI,   MVC_ARI, MVI_ARI, MVCI_ARI

procedure write (x: Type2);

Type2:   *integer, boolean, char, string, real, complex, interval, cinterval, rvector, cvector, ivector, civector, rmatrix, cmatrix, imatrix, cimatrix*

Purpose: Corresponds to *write(output*, x).

writeln

procedure writeln (**var** f: text);

Purpose: Termination of an output line by writing the end-of-line character. It is an error if *f* is undefined or if *f* is not in writing mode.

procedure writeln;

Purpose: Corresponds to *writeln (output)*.

procedure writeln (**var** f: text; x: Type2);

Type2:   *integer, boolean, char, string, real, complex, interval, cinterval, rvector, cvector, ivector, civector, rmatrix, cmatrix, imatrix, cimatrix*

Purpose: Corresponds to *write(f*, x) followed by *writeln(f)*.

procedure writeln (x: Type2);

Type2:   *integer, boolean, char, string, real, complex, interval, cinterval, rvector, cvector, ivector, civector, rmatrix, cmatrix, imatrix, cimatrix*

Purpose: Corresponds to *write(x)* followed by *writeln*.

B.8 #-Expressions

B.8.1 Real and Complex #-Expressions

Syntax: #-Symbol (Exact Expression)

#-Symbol	Result Type	Summands Permitted in the Exact Expression
#	dotprecision	• variables, constants, and special function calls of type *integer, real,* or *dotprecision* • products of type *integer* or *real* • scalar products of type *real*
#* #< #>	real	• variables, constants, and special function calls of type *integer, real,* or *dotprecision* • products of type *integer* or *real* • scalar products of type *real*
	complex	• variables, constants, and special function calls of type *integer, real, complex,* or *dotprecision* • products of type *integer, real,* or *complex* • scalar products of type *real* or *complex*
	rvector	• variables and special function calls of type *rvector* • products of type *rvector* (e.g. *rmatrix* ∗ *rvector, real* ∗ *rvector* etc.)
	cvector	• variables and special function calls of type *rvector* or *cvector* • products of type *rvector* or *cvector* (e.g. *cmatrix* ∗ *rvector, real* ∗ *cvector* etc.)
	rmatrix	• variables and special function calls of type *rmatrix* • products of type *rmatrix*
	cmatrix	• variables and special function calls of type *rmatrix* or *cmatrix* • products of type *rmatrix* or *cmatrix*

"Special function calls" are the calls of the functions *compl, re, im, conj, intval, inf, sup, id, transp, herm,* and the type converting functions *rvector, cvector, ivector,* and *civector.*

B.8.2 Real and Complex Interval #-Expressions

Syntax: ## (Exact Expression)

#-Symbol	Result Type	Summands Permitted in the Exact Expression
##	interval	• variables, constants, and special function calls of type *integer*, *real*, *interval*, or *dotprecision* • products of type *integer*, *real*, or *interval* • scalar products of type *real* or *interval*
	cinterval	• variables, constants, and special function calls of type *integer*, *real*, *complex*, *interval*, *cinterval*, or *dotprecision* • products of type *integer*, *real*, *complex*, *interval*, or *cinterval* • scalar products of type *real*, *complex*, *interval*, or *cinterval*
	ivector	• variables and special function calls of type *rvector* or *ivector* • products of type *rvector* or *ivector*
	civector	• variables and special function calls of type *rvector*, *cvector*, *ivector*, or *civector* • products of type *rvector*, *cvector*, *ivector*, or *civector*
	imatrix	• variables and special function calls of type *rmatrix* or *imatrix* • products of type *rmatrix* or *imatrix*
	cimatrix	• variables and special function calls of type *rmatrix*, *cmatrix*, *imatrix*, or *cimatrix* • products of type *rmatrix*, *cmatrix*, *imatrix*, or *cimatrix*

"Special function calls" are the calls of the functions *compl, re, im, conj, intval, inf, sup, id, transp, herm*, and the type converting functions *rvector, cvector, ivector*, and *civector*.

Bibliography

[1] Alefeld, G. and Herzberger, J.: *Einführung in die Intervallrechnung*. Bibliographisches Institut, Mannheim, 1974.

[2] Alefeld, G. and Herzberger, J.: *Introduction to Interval Computations*. Academic Press, New York, 1983.

[3] American National Standards Institute / Institute of Electrical and Electronic Engineers: *A Standard for Binary Floating-Point Arithmetic*. ANSI/IEEE Std. 754-1985, New York, 1985.

[4] Bauch, H., Jahn, K.-U., Oelschlägel, D., Süsse, H., and Wiebigke, V.: *Intervallmathematik*. Teubner, Leipzig, 1987.

[5] Bleher, J. H., Rump, S. M., Kulisch, U., Metzger, M., Ullrich, Ch., and Walter, W.: *FORTRAN-SC: A Study of a FORTRAN Extension for Engineering/Scientific Computation with Access to ACRITH*. Computing **39**, pp 93-110, 1987.

[6] Bohlender, G., Rall, L., Ullrich, Ch., and Wolff von Gudenberg, J.: *PASCAL-SC: A Computer Language for Scientific Computation*. Academic Press, New York, 1987.

[7] Bohlender, G., Rall, L., Ullrich, Ch., and Wolff von Gudenberg, J.: *PASCAL-SC – Wirkungsvoll programmieren, kontrolliert rechnen*. Bibliographisches Institut, Mannheim, 1986.

[8] Böhm, H.: *Auswertung arithmetischer Ausdrücke mit maximaler Genauigkeit*. In: Kulisch, U. and Ullrich, Ch. (Eds.): *Wissenschaftliches Rechnen und Programmiersprachen*. Berichte des German Chapter of the ACM, Band 10, pp 175-184, Teubner, Stuttgart, 1982.

[9] British Standards Institution: *Specification for Computer Programming Language PASCAL*. BS 6192:1982, UDC 681.3.06, PASCAL:519.682, London, 1982.

[10] IBM *High-Accuracy Arithmetic Subroutine Library* (ACRITH). General Information Manual, GC 33-6163-02, 3rd Edition, 1986.

[11] IBM *High-Accuracy Arithmetic Subroutine Library* (ACRITH). Program Description and User's Guide, SC 33-6164-02, 3rd Edition, 1986.

[12] IBM *High Accuracy Arithmetic – Extended Scientific Computation* (ACRITH-XSC), Reference. SC 33-6462-00, IBM Corp., 1990.

[13] Jensen, K. and Wirth, N.: *PASCAL User Manual and Report*. ISO PASCAL Standard, 3rd ed., Springer, Berlin, 1985.

[14] Kaucher, E., Klatte, R., and Ullrich, Ch.: *Programmiersprachen im Griff – Band 2: PASCAL*. Bibliographisches Institut, Mannheim, 1981.

[15] Kaucher, E., Klatte, R., Ullrich, Ch., and Wolff von Gudenberg, J.: *Programmiersprachen im Griff – Band 4: ADA*. Bibliographisches Institut, Mannheim, 1983.

[16] Kaucher, E., Kulisch, U., and Ullrich, Ch. (Eds.): *Computer Arithmetic – Scientific Computation and Programming Languages*. Teubner, Stuttgart, 1987.

[17] Kaucher, E. and Miranker, W. L.: *Self-Validating Numerics for Function Space Problems*. Academic Press, New York, 1984.

[18] Kaucher, E. and Rump, S. M.: *E-Methods for Fixed Point Equation $f(x) = x$*. Computing **28**, pp 31-42, 1982.

[19] Kießling, I., Lowes, M., and Paulik, A.: *Genaue Rechnerarithmetik - Intervallrechnung und Programmieren mit PASCAL-SC*. Teubner, Stuttgart, 1988.

[20] Kirchner, R. and Kulisch, U.: *Accurate Arithmetic for Vector Processors*. Journal of Parallel and Distributed Computing **5**, pp 250-270, 1988.

[21] Klatte, R., Kulisch, U., Neaga, M., Ratz, D., and Ullrich, Ch.: *PASCAL–XSC – Sprachbeschreibung mit Beispielen*. Springer-Verlag, Heidelberg, 1991.

[22] Klatte, R. and Ullrich, Ch.: *Programmiersprachen im Griff – Band 9: MODULA-2*. Bibliographisches Institut, Mannheim, 1988.

[23] Knuth, D. E.: *The Art of Computer Programming*. Vol. 2: *Seminumerical Algorithms*. Addison Wesley, Reading, Massachusetts., 1981.

[24] Kulisch, U.: *Grundlagen des Numerischen Rechnens – Mathematische Begründung der Rechnerarithmetik*. Reihe Informatik, Band 19, Bibliographisches Institut, Mannheim, 1976.

[25] Kulisch, U. (Ed.): *PASCAL-SC: A PASCAL Extension for Scientific Computation*, Information Manual and Floppy Disks, Version ATARI ST. Teubner, Stuttgart, 1987.

[26] Kulisch, U. (Ed.): *PASCAL-SC: A PASCAL Extension for Scientific Computation*, Information Manual and Floppy Disks, Version IBM PC/AT (DOS). Teubner, Stuttgart, 1987.

[27] Kulisch, U. (Ed.): *Wissenschaftliches Rechnen mit Ergebnisverifikation – Eine Einführung*. Akademie Verlag, Ost-Berlin, Vieweg, Wiesbaden, 1989.

[28] Kulisch, U. and Miranker, W. L.: *Computer Arithmetic in Theory and Practice*. Academic Press, New York, 1981.

[29] Kulisch, U. and Miranker, W. L.: *The Arithmetic of the Digital Computer: A New Approach*. SIAM Review, Vol. 28, No. 1, pp 1-140, 1986.

[30] Kulisch, U. and Miranker, W. L. (Eds.): *A New Approach to Scientific Computation*. Academic Press, New York, 1983.

[31] Kulisch, U. and Stetter, H. J. (Eds.): *Scientific Computation with Automatic Result Verification*. Computing Suppl. **6**, Springer-Verlag, Vienna, 1988.

[32] Kulisch, U. and Ullrich, Ch. (Eds.): *Wissenschaftliches Rechnen und Programmiersprachen*. Berichte des German Chapter of the ACM, Band 10, Teubner, Stuttgart, 1982.

[33] Mayer, G.: *Grundbegriffe der Intervallrechnung*. In: Kulisch, U. (Ed.): *Wissenschaftliches Rechnen mit Ergebnisverifikation – Eine Einführung*, pp 101-118, Akademie Verlag, Ost-Berlin, Vieweg, Wiesbaden, 1989.

[34] Moore, R. E.: *Interval Analysis*. Prentice Hall, Engelwood Cliffs, New Jersey, 1966.

[35] Moore, R. E.: *Methods and Applications of Interval Analysis*. SIAM, Philadelphia, Pensylvania, 1979.

[36] Moore, R. E. (Ed.): *Reliability in Computing: The Role of Interval Methods in Scientific Computations*. Academic Press, New York, 1988.

[37] Neaga, M.: *PASCAL-SC – Eine PASCAL-Erweiterung für wissenschaftliches Rechnen*. In: Kulisch, U. (Ed.): *Wissenschaftliches Rechnen mit Ergebnisverifikation – Eine Einführung*, pp 69-84. Akademie Verlag, Ost-Berlin, Vieweg, Wiesbaden, 1989.

[38] Neumaier, A.: *Interval Methods for Systems of Equations*. Cambridge University Press, Cambridge, 1990.

[39] Nickel, K. (Ed.): *Interval Mathematics*. Proceedings of the International Symposium, Karlsruhe 1975, Springer-Verlag, Vienna, 1975.

[40] Nickel, K. (Ed.): *Interval Mathematics 1980*. Proceedings of the International Symposium, Freiburg 1980, Academic Press, New York, 1980.

[41] Nickel, K. (Ed.): *Interval Mathematics 1985*. Proceedings of the International Symposium, Freiburg 1985, Springer-Verlag, Vienna, 1986.

[42] Rall, L. B.: *Automatic Differentiation, Techniques and Applications*. Lecture Notes in Computer Science, No. 120, Springer, Berlin, 1981.

[43] Ratschek, H. and Rokne, J.: *Computer Methods for the Range of Functions*. Ellis Horwood Limited, Chichester, 1984.

[44] Rump, S. M.: *Lösung linearer und nichtlinearer Gleichungssysteme mit maximaler Genauigkeit*. In: Kulisch, U. and Ullrich, Ch. (Eds.): *Wissenschaftliches Rechnen und Programmiersprachen*. Berichte des German Chapter of the ACM, Band 10, pp 147-174, Teubner, Stuttgart, 1982.

[45] Rump, S. M.: *Wie zuverlässig sind die Ergebnisse unserer Rechenanlagen*. In: Jahrbuch Überblicke Mathematik, Bibliographisches Institut, Mannheim, 1983.

[46] Rump, S. M.: *Solving Algebraic Problems with High Accuracy*. In: Kulisch, U. and Miranker, W. L. (Eds.): *A New Approach to Scientific Computation*, pp 51-120. Academic Press, New York, 1983.

[47] Stoer, J. and Bulirsch, R.: *Introduction to Numerical Analysis*. Springer-Verlag, New York, 1980.